Lecture Notes in Networks and Systems 1776

The series "Lecture Notes in Networks and Systems" publishes the latest developments in Networks and Systems—quickly, informally and with high quality. Original research reported in proceedings and post-proceedings represents the core of LNNS.

Volumes published in LNNS embrace all aspects and subfields of, as well as new challenges in, Networks and Systems.

The series contains proceedings and edited volumes in systems and networks, spanning the areas of Cyber-Physical Systems, Autonomous Systems, Sensor Networks, Control Systems, Energy Systems, Automotive Systems, Biological Systems, Vehicular Networking and Connected Vehicles, Aerospace Systems, Automation, Manufacturing, Smart Grids, Nonlinear Systems, Power Systems, Robotics, Social Systems, Economic Systems and other. Of particular value to both the contributors and the readership are the short publication timeframe and the world-wide distribution and exposure which enable both a wide and rapid dissemination of research output.

The series covers the theory, applications, and perspectives on the state of the art and future developments relevant to systems and networks, decision making, control, complex processes and related areas, as embedded in the fields of interdisciplinary and applied sciences, engineering, computer science, physics, economics, social, and life sciences, as well as the paradigms and methodologies behind them.

Indexed by SCOPUS, EI Compendex, INSPEC, WTI Frankfurt eG, zbMATH, SCImago.

All books published in the series are submitted for consideration in Web of Science.

For proposals from Asia please contact Aninda Bose (aninda.bose@springer.com).

Manuel Rodrigues · Ichiro Satoh ·
Pablo Chamoso · Ricardo S. Alonso ·
Paulo Novais · Antoine Nongaillard
Editors

Ambient Intelligence – Software and Applications – 16th International Symposium on Ambient Intelligence

Springer

Editors
Manuel Rodrigues
ALGORITMI Centre/LASI Departamento de Informática
Universidade do Minho
Braga, Portugal

Pablo Chamoso
Bioinformatics, Intelligent Systems and Educational Technology (BISITE) Research Group
University of Salamanca
Salamanca, Salamanca, Spain

Paulo Novais
ALGORITMI Centre/LASI Departamento de Informática
Universidade do Minho
Braga, Portugal

Ichiro Satoh
National Institute of Informatics
Chiyoda, Tokyo, Japan

Ricardo S. Alonso
Escuela Superior de Ingeniería y Tecn.
Universidad Internacional de La Rioja
Logroño, Spain

Antoine Nongaillard
University of Lille
Villeneuve-d'Ascq, France

ISSN 2367-3370 ISSN 2367-3389 (electronic)
Lecture Notes in Networks and Systems
ISBN 978-3-032-14137-8 ISBN 978-3-032-14138-5 (eBook)
https://doi.org/10.1007/978-3-032-14138-5

This Springer imprint is published by the registered company Springer Nature Switzerland AG
The registered company address is: Gewerbestrasse 11, 6330 Cham, Switzerland

Preface

Ambient Intelligence (AmI) is a paradigm in distributed computing and artificial intelligence that envisions the seamless integration of technology into everyday environments. The core idea is to create adaptive, intelligent spaces that can sense, interpret, and respond to human presence, needs, and behavior without requiring explicit commands. These environments rely on networks of embedded devices, sensors, and systems that function unobtrusively, all while upholding individuals' digital rights and privacy. Unlike traditional automation, AmI emphasizes contextual awareness and user-centric design, aiming to enhance daily life through personalized, anticipatory interactions. Recent advancements in foundational AI models, such as Large Language Models (LLMs) and Large Multimodal Models (LMMs), as well as progress in quantum computing for molecular simulations and optimization tasks, are significantly expanding the capabilities of AmI systems.

The applications of Ambient Intelligence span across diverse sectors, notably in healthcare, smart living, and urban infrastructure. In healthcare and social care, AmI offers transformative solutions for continuous patient monitoring, early detection of cognitive diseases, and assistance for elderly or dependent individuals through natural language interfaces and ambient sensing. It can support visually impaired individuals by providing real-time guidance in both public and private spaces. Within smart homes and buildings, AmI systems can autonomously regulate lighting, temperature, and energy use based on user habits, while also enhancing security through intelligent surveillance. In the context of smart cities, AmI can contribute to public safety by detecting aggressive behaviors using video analytics and monitoring environmental parameters like water or air quality through distributed sensor networks.

Overall, Ambient Intelligence aims to create responsible, context-aware ecosystems that blend technology seamlessly into human life, enriching user experiences while ensuring ethical data practices. The convergence of AI, ubiquitous computing, sensor technology, and quantum innovation positions AmI as a critical component of the digital future. As these systems evolve, they hold the potential to redefine how we live, work, and interact—moving toward environments that are not only smarter but also more humane, inclusive, and sustainable.

In this regard, ISAmI is the *International Symposium on Ambient Intelligence*, which seeks to bring together scientists working in all areas related to AmI, as can be seen in the articles included in the conference. This year, we are proud to tell that ISAmI is in its 16th edition, including several papers focused on the different key aspects of the AmI. Interested researchers can find works on advanced applications of Context-Aware Intelligence and Sensing for Smart Environments; AI-Driven Applications and Decision Support in Clinical and Personalized Healthcare; as well as Advanced AI and Interactive Systems for Emerging Technological Domains, such as Large Language Models, Extended Reality or 5G/6G networks.

In total, 12 articles have been selected from a total of 18 submissions from many different countries (including Brazil, Greece, Japan, Palestine, Portugal, Spain, and the United Kingdom).

As in previous editions, there are special issues in JCR-ranked journals such as the *International Journal of Interactive Multimedia and Artificial Intelligence; the Logic Journal of the IGPL; the Advances in Distributed Computing and Artificial Intelligence Journal; Future Internet; Sensors; as well as Systems.*

The symposium is organized by the Université of Lille (France), the Universidade do Minho (Portugal), the National Institute of Informatics (Japan), the Università degli Studi di Genova (Italy), the University of Granada (Spain), the AIR Institute (Spain), and the University de Salamanca (Spain).

This edition of the conference was hosted by the SMAC Team of the CRIStAL Lab of the University of Lille (France) from June 25th to 27th, 2025. We would like to thank all the contributing authors, the members of the Program Committee for their hard and highly valuable work. Without their efforts, it would have been impossible to put together such an interesting and actual program.

Finally, we appreciate the efforts of local organizers, for hosting and organizing these events, without which ISAmI could not have taken place.

Manuel Rodrigues
Ichiro Satoh
Pablo Chamoso
Ricardo S. Alonso
Paulo Novais
Antoine Nongaillard

Organization

General chair

Paulo Novais	University of Minho, Portugal

Program Committee chairs

Ichiro Satoh	National Institute of Informatics, Japan
Manuel Rodrigues	University of Minho, Portugal

Advisory Board

Gianni Vercelli	Università di Genova, Italy
Miguel J. Hornos	University of Granada, Spain
Ricardo S. Alonso	AIR Institute, Spain
Pablo Chamoso	University of Salamanca, Spain

Organizing committee

Philippe Mathieu (Chair)	University of Lille (France)
Anne-Cécile Caron	University of Lille (France)
Antoine Nongaillard	University of Lille (France)
Jarod Vanderlynden	University of Lille (France)
Jules Bompard	University of Lille (France)
Erwan Martin	University of Lille (France)

Program Committee

Murad Al-Rajab	Abu Dhabi University (United Arab Emirates)
Cesar Analide	University of Minho (Portugal)
Masayoshi Aritsugi	Kumamoto University (Japan)
Sara Balderas-Díaz	University of Cádiz (Spain)
Chintan Bhatt	Pandit Deendayal Energy University (India)
Valérie Camps	University of Toulouse III - IRIT France (France)
Javier Carbo	Univ. Carlos III of Madrid (Spain)

Fabio Cassano	Università degli Studi di Bari Aldo Moro (Italy)
Jose Carlos Castillo Montoya	Universidad Carlos III de Madrid (Spain)
Alvaro Castro-Gonzalez	Universidad Carlos III de Madrid (Spain)
Silvio Cesar Cazella	UFCSPA (Brazil)
Pablo Chamoso	University of Salamanca (Spain)
Stefano Chessa	University of Pisa (Italy)
Stéphanie Combettes	University of Toulouse III - IRIT France (France)
Antonio Curci	University of Bari Aldo Moro (Italy)
Luis de La Fuente Valentín	Universidad Internacional de La Rioja (Spain)
Patricio Domingues	ESTG - Leiria (Portugal)
Dalila Durães	Universidade do Minho (Portugal)
Luiz Faria	Polytechnic of Porto (Portugal)
Florentino Fdez-Riverola	University of Vigo (Spain)
Antonio Fernández-Caballero	Universidad de Castilla-La Mancha (Spain)
Lino Figueiredo	ISEP (Portugal)
Oscar Garcia	Universidad Internacional de La Rioja (Spain)
Jose Garcia-Alonso	University of Extremadura (Spain)
Sylvain Giroux	Université de Sherbrooke (Canada)
Luis Gomes	Universidade dos Açores (Portugal)
Marco Gomes	University of Minho (Portugal)
Jose Luis Gomez-Sirvent	UCLM (Spain)
Celestino Goncalves	Instituto Politecnico da Guarda (Portugal)
Alfonso González Briones	University of Salamanca (Spain)
Sérgio Gonçalves	University of Minho (Portugal)
Junzhong Gu	East China Normal University (China)
Esteban Guerrero	Umeå university (Sweden)
Rosa Graciela Guerrero	Universidad de las Fuerzas Armadas (Ecuador)
Gabriel Guerrero-Contreras	University of Cadiz (Spain)
Miguel Guimarães	INESC TEC (Portugal)
Guillermo Hernández	Universidad de Salamanca (Spain)
Miguel J. Hornos	University of Granada (Spain)
Manal Jammal	Air Institute (Spain)
Vicente Julian	Universitat Politècnica de València (Spain)
Jason Jung	Chung-Ang University (South Korea)
Alexander Kocian	University of Pisa (Italy)
Igor Kotenko	St. Petersburg Federal Research Center of the Russian Academy of Sciences (Russia)
Paula Lamo	Universidad Internacional de La Rioja (Spain)
Tiancheng Li	Northwestern Polytechnical University (China)
Raúl López Blanco	Universidad de Salamanca (Spain)
José Machado	University of Minho (Portugal)
João Paulo Magalhaes	Porto Polytechnic Institute (Portugal)

Sadouanouan Malo	University Nazi Boni (Burkina Faso)
Arturo Martinez-Rodrigo	University of Castilla-LA Mancha (Spain)
Constantino Martins	Polytechnic of Porto (Portugal)
Paulo Matos	Instituto Politécnico de Bragança (Portugal)
Rene Meier	Lucerne University of Applied Sciences (Switzerland)
Pedro Melo-Pinto	CITAB - UTAD (Portugal)
Israel Mendonça	Kumamoto University (Brazil)
Yeray Mezquita	Universidad de Salamanca (Spain)
Andres Muñoz	Universidad de Cádiz (Spain)
Federico Muñoz Babiano	UNIR (Spain)
Sergio Márquez Sánchez	AIR institute (Spain)
Tatsuo Nakajima	Waseda University (Japan)
Elena Navarro	University of Castilla-La Mancha (Spain)
Suhail Odeh	Bethlehem University (Palestine)
Andrei Olaru	University Politehnica of Bucharest (Romania)
Pedro Filipe Oliveira	Instituto Politécnico de Bragança (Portugal)
Jaderick Pabico	University of the Philippines Los Banos (Philippines)
Hugo Peixoto	University of Minho (Portugal)
António Pinto	ESTGP Porto (Portugal)
Tiago Pinto	Universidade de Trás-os-Montes e Alto Douro/INESC-TEC (Portugal)
Francisco Pinto-Santos	University of Salamanca (Spain)
João Ramos	Polytechnic Institute of Porto (Portugal)
José Carlos Ribeiro	Polytechnic Institute of Leiria (Portugal)
Guillermo Rivas	AIR Institute (Spain)
Manuel Rodrigues	Minho University (Portugal)
Nuno Rodrigues	Politécnico de Leiria (Portugal)
Sara Rodríguez	University of Salamanca (Spain)
Iván Rodríguez Conde	Universidade do Vigo (Spain)
Altino Sampaio	Instituto Politécnico do Porto (Portugal)
Roberto Sanchez Reolid	Universidad de Castilla-La Mancha (Spain)
Ricardo Santos	Polytechnic Institute of Porto (Portugal)
Ichiro Satoh	National Institute of Informatics (Japan)
Yaqin Shaheen	Palestine Technical University (Palestine)
Fernando Silva	Polytechnic of Leiria (Portugal)
Fábio Silva	University of Minho (Portugal)
Diego Valdeolmillos Villaverde	University of Salamanca (Spain)
Radu-Daniel Vatavu	University Stefan cel Mare of Suceava (Romania)
Fatos Xhafa	Universitat Politècnica de Catalunya Barcelona (Spain)

ISAmI'25 Organizers

Organizing Committee

Université de Lille

Scientific Organizers

Universidade do Minho

NII

UNIVERSITÀ DEGLI STUDI DI GENOVA

UNIVERSIDAD DE GRANADA

AIR INSTITUTE

VNIVERSIDAD D SALAMANCA

Contents

Development of a Graphical Application in Python for the Clinical Study of Patients with Hyperhidrosis

Tiago Rocha Pereira[1,4](✉), Fatima Carvalho[2,4], Joaquim Gonçalves[3,4], and Constantino Martins[1,4]

[1] School of Engineering of Polytechnic Institute of Porto (ISEP-IPP), Porto, Portugal
tiago.pereira@eu.ipp.pt, acm@isep.ipp.pt
[2] Centro Hospitalar do Porto, Porto, Portugal
[3] Instituto Politécnico do Cávado e do Ave, Barcelos, Portugal
jgoncalves@ipca.pt
[4] GECAD - Research Group on Intelligent Engineering and Computing for Advanced Innovation and Development, Porto, Portugal

Abstract. Hyperhidrosis (HH) is a condition characterized by excessive sweating beyond physiological needs, affecting patients' quality of life. This study introduces an innovative technological solution focused on the development of a graphical application to support the clinical study of hyperhidrosis, a condition characterized by excessive sweating. The application, developed using Python, Raspberry Pi, and temperature sensors, enables real-time monitoring of thermal variations. This system provides a novel approach to clinical data collection, specifically focusing on how biometric data is measured and how visual and auditory alerts are generated. The application's design and implementation, highlighting the integration of PyQt4 and Matplotlib for a Python-based GUI, are the primary focus of this paper.

Keywords: Hyperhidrosis · Python · application · temperature · humidity · graphical · sensors · Single Board Computer

1 Introduction

Primary hyperhidrosis (HH) is a chronic condition characterized by excessive sweating that exceeds the physiological response. This excessive perspiration primarily affects specific areas of the body, such as the palms, soles, axillae, and face, significantly impacting patients' quality of life.

The prevalence of idiopathic hyperhidrosis is not well established, with estimates ranging from 1.5% to 5.8%. Despite its impact, the pathophysiology of hyperhidrosis remains poorly understood. In the absence of an anatomical cause, hyperhidrosis is considered a result of a complex dysfunction of the autonomic nervous system.

Previous studies have used skin temperature to evaluate the effectiveness of surgical procedures for treating hyperhidrosis. However, this paper focuses on the development

M. Rodrigues et al. (Eds.): ISAmI 2025, LNNS 1776, pp. 1–11, 2026.
https://doi.org/10.1007/978-3-032-14138-5_1

of a novel application for monitoring skin temperature responses, detailing the system's architecture and software implementation.

2 Methods

2.1 Ethical Approval

This study was approved by the Ethics Committee of the Centro Hospitalar Universitário do Porto (CHUP) under reference number 2018.140 (121-DEFI/120-CES).

2.2 Participants

A total of 37 patients diagnosed with primary hyperhidrosis and 16 age- and sex-matched controls participated in the study. Participants were selected from the outpatient pediatric surgery clinic of Hospital Santo António (Porto, Portugal) between January 2017 and July 2019. The diagnosis of primary hyperhidrosis was based on reported symptoms of persistent, excessive sweating that limits routine activities, affecting areas such as the palms, soles, and axillae.

Written informed consent was obtained from all participants, and the study adhered to ethical guidelines. The participants underwent thermography procedures as outlined in the Glamorgan Protocol.

3 Prototype Development

3.1 Environmental Setup

A controlled physical environment was designed to regulate ambient temperature and humidity. The environment consisted of a room where both temperature and humidity could be adjusted and monitored in real-time. This setup was vital for simulating the thermal conditions under which the patients would be assessed, ensuring that the variables were consistent across all tests (Figs. 1 and 2).

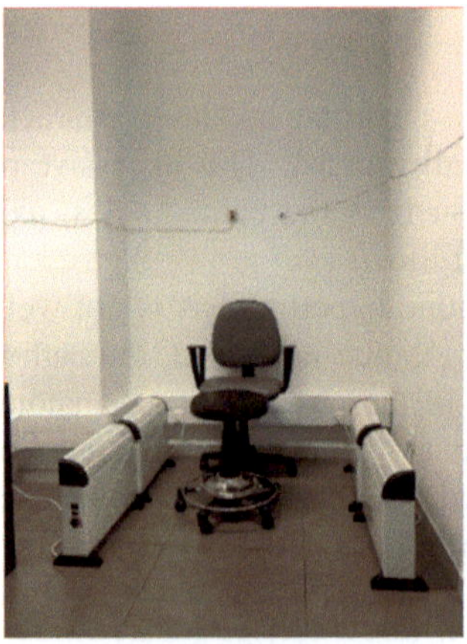

Fig. 1. Room setup

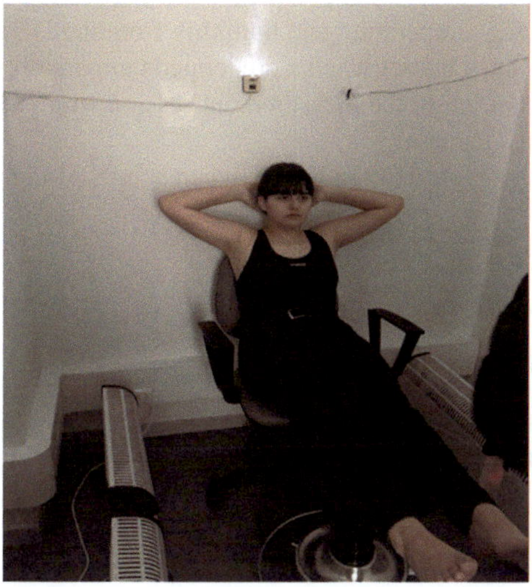

Fig. 2. Measurement session (thermometer on top of the test subject)

3.2 System Architecture

The heart of the monitoring system is a Raspberry Pi Single-Board Computer (SBC), chosen for its versatility, energy efficiency, and compact size. The Raspberry Pi integrates with the DHT22/AM2302 temperature and humidity sensor, which records the temperature variations of the environment. This data is processed using Python and displayed in real-time via a graphical user interface (GUI) developed using PyQt4. The application also leverages Matplotlib for generating real-time graphical plots of temperature variation over time.

The Raspberry Pi was chosen due to its low cost and ease of integration with various sensors and components through its General Purpose Input/Output (GPIO) pins. This allowed for simple wiring and seamless connection of the DHT22 sensor to the Pi. Additionally, Python was used because of its simplicity and extensive libraries, making it the ideal language for rapid prototyping and future maintenance.

3.3 System Components

The main components of the system are based in easy to find, robust, well documented, yet affordable hardware components:

- **Raspberry Pi**: The Raspberry Pi 3 Model B was used for this prototype, which includes a quad-core ARM Cortex-A53 CPU, 1 GB of RAM, and onboard Wi-Fi connectivity [1].
- **DHT22/AM2302 Sensor**: This sensor was selected for its accuracy and reliability in measuring temperature and humidity. It has a wide measurement range (temperature: −40 to 80 °C; humidity: 0–100%) and a fast response time [2].
- **GPIO Pins**: Several GPIO pins on the Raspberry Pi were used for various control functions: [3]

- GPIO 4 for reading temperature and humidity from the DHT22 sensor.
- GPIO 21 to activate a buzzer when a predefined temperature threshold is reached.
- GPIO 20 to start or pause the measurement session while taking images with the FLIR E60 thermal camera.
- GPIO 23 to activate an LED indicator when a critical temperature threshold is reached.

3.4 Software and Libraries

The software that powers the system was developed in Python, using the following libraries:

1. **GPIO Zero:** This high-level library was used for controlling the Raspberry Pi's GPIO pins, including reading sensor data and activating alerts such as the buzzer and LED. The library abstracts much of the low-level code, simplifying interactions with hardware components [3].
2. **Adafruit_Python_DHT:** This library was used to interface with the DHT22/AM2302 sensor to read temperature and humidity data. It allows for simple and efficient communication with the sensor, enabling real-time data collection [4].
3. **PiGPIO:** For more precise timing and better control over GPIO operations, PiGPIO was utilized, particularly for interacting with the GPIO pins when precise timing was necessary [5].
4. **PyQt4:** This library was used to develop the graphical user interface (GUI). The GUI was designed to display the real-time data from the sensors and to provide visual alerts when the temperature exceeds predefined thresholds.
5. **Matplotlib:** Matplotlib was used for visualizing the temperature variations over time. The real-time data was plotted in the GUI, providing a graphical representation of the environmental conditions and allowing users to track temperature changes during the monitoring process.

3.5 System Workflow

The workflow between the system components was designed with simplicity in mind, for maintainability, consisting in a continuous loop collecting samples from the DHT22 sensor, discarding out-of-band readings caused by interference, and calculating the medium temperature of the valid samples.

Then, when the medium temperature hits a predefined threshold, an alert is triggered, so that the test subject's body temperature can be collected using a FLIR thermal camera.

1. **Data Collection:** The DHT22 sensor continuously measures the temperature and humidity in the environment [6]. The data is collected at regular intervals and sent to the Raspberry Pi.
2. **Data Processing:** The temperature data is processed by the Python application running on the Raspberry Pi. The application calculates whether the temperature exceeds predefined thresholds (20 °C, 24 °C, 28 °C, and 32 °C).
3. **Alert System:** If the temperature exceeds a threshold, the system triggers an alert:

   - Visual Alert: A message is displayed in the GUI, indicating the current temperature and the threshold that has been surpassed.
   - Auditory Alert: The buzzer connected to GPIO 21 is activated to provide an audible notification.

4. **Graphical Visualization:** The temperature data is plotted in real-time using Matplotlib. The user can observe the temperature trend over time through the GUI, with the ability to zoom in on specific time periods for further analysis.
5. **Data Logging:** The system logs the temperature data with a timestamp (in the format Year-Month-Day_Hour:Minute:Second) to ensure accurate records for further analysis. The data is saved to a text file for later use, and can also be exported to CSV format for detailed statistical analysis.

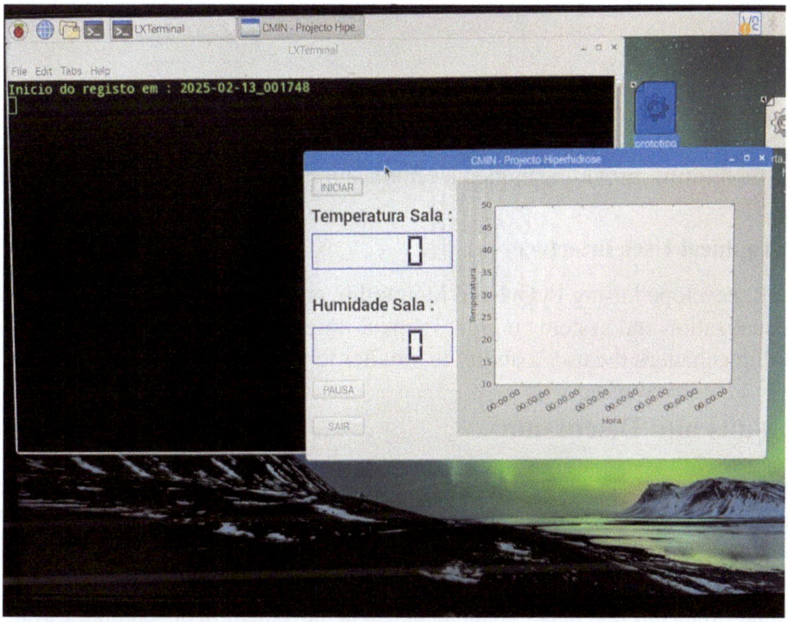

Fig. 3. The application's GUI developed using Python, Qt4 and Mathplot Library. As soon as the application opens, all information is set to default "0"

3.6 System User Interface (UI)

The graphical user interface (GUI) Fig. 3 is the primary means through which users interact with the system. Built with PyQt4, it provides a simple and intuitive layout for monitoring temperature data and receiving alerts. The GUI features:

- **Real-time data display:** Current temperature and humidity are displayed on LCD-like counters.
- **Graphical temperature plot:** Real-time temperature data is visualized using Matplotlib.
- **Alert notifications:** Visual and auditory alerts when the temperature exceeds the predefined thresholds.

Control buttons: Buttons to start/stop measurements and save data.

4 Process

This section details the process and usage of the prototype, focusing on how it measures biometric data and generates alerts.

4.1 Data Acquisition and Processing

The application successfully acquires temperature and humidity data from the DHT22 sensor and processes it in real-time. The system's ability to filter out-of-band readings ensures the accuracy of the data. The calculation of the median temperature provides a stable metric for triggering alerts.

4.2 Alert Generation

The application effectively generates both visual and auditory alerts when predefined temperature thresholds are exceeded. This real-time alerting capability is crucial for clinical monitoring, providing immediate notification of significant temperature changes.

4.3 Graphical User Interface

The GUI, developed using PyQt4 and Matplotlib, provides a user-friendly interface for data visualization and system control. The real-time plotting of temperature data using Matplotlib enhances the user's ability to monitor temperature trends.

5 Results and Discussion

This study demonstrates the development of a novel Python-based graphical application for real-time monitoring of thermal variations. The application's innovative use of PyQt4 and Matplotlib for data presentation and the effective integration of hardware and software components are key contributions. While the clinical implications are significant, the primary focus of this paper is the technical achievement in developing a system for accurate and real-time temperature monitoring and alerting.

5.1 Real-Time Monitoring and Alerting System

The Raspberry Pi-based application proved effective in monitoring and alerting in real-time. The system displayed the temperature data visually in the GUI, with a dynamic graph showing the temperature progression over time. When the temperature exceeded the predefined threshold of 28 °C, the system activated a visual and auditory alert, notifying the user of the change. This feature was instrumental in providing immediate feedback on the patient's condition during the test.

Additionally, the system logged the data for post-experiment analysis, recording the temperature values and time stamps. The system's ability to record and visualize the data in real-time provided useful insights into the autonomic dysfunction present in the HH patients.

5.2 Future Directions

The potential applications of this system extend beyond the research context. In the future, this system could be integrated into clinical practice for routine monitoring of HH patients, allowing for continuous, non-invasive tracking of thermoregulatory responses. This would provide clinicians with valuable data to tailor treatment plans and evaluate patient progress. Additionally, the system could be expanded to monitor other symptoms of autonomic dysfunction, such as changes in skin perfusion or sweating patterns.

Future iterations of the system could also benefit from incorporating machine learning algorithms to analyse the collected data. By using artificial intelligence to predict trends and identify patterns in real-time, the system could assist healthcare providers in making more informed decisions regarding treatment and management.

6 Conclusions

6.1 HH Study Conclusion

Thermal images were collected from 37 patients with hyperhidrosis and 16 controls using a FLIR E60 thermal camera. The images were analysed to assess temperature differences across various body areas.

The records were made according to the established protocol and the assumptions previously foreseen, the data refer to the areas of clinical impact (palms, soles of the feet and armpits) and the core body temperature recorded in the internal epicorners of both palpebral fissures. The areas studied were chosen taking into account the influence of dermatomes (Fig. 4), angiosomes (Fig. 5) and the distribution map of the sweat glands (Fig. 6) already known.

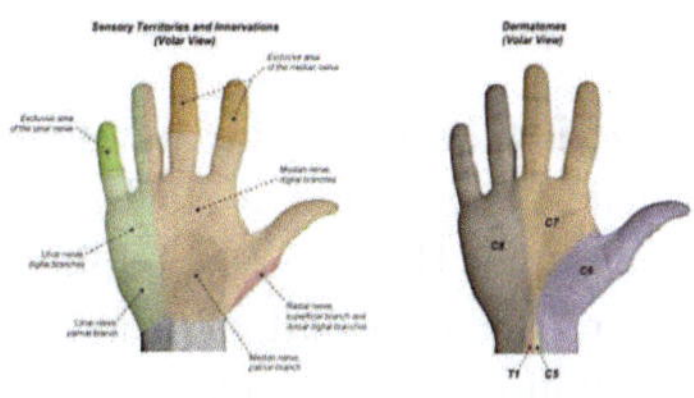

Fig. 4. Dermatomes palmares

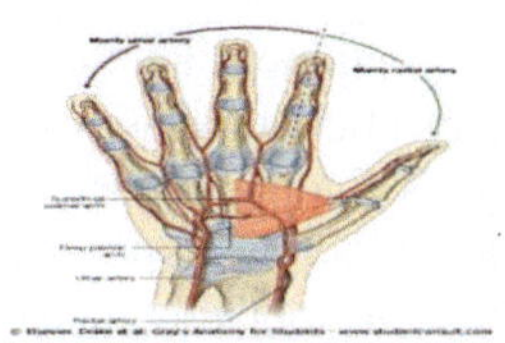

Fig. 5. Angiosomes palmares

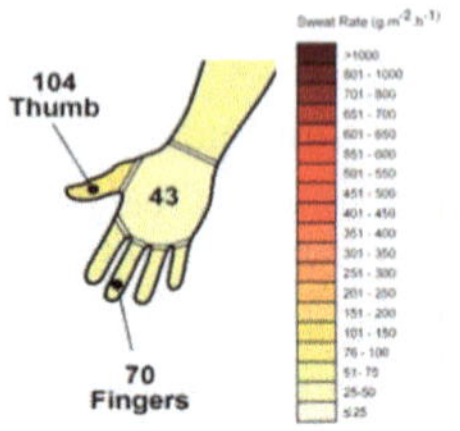

Fig. 6. Sweat glands

The analysis for all groups is made using the values of the body thermal records reached during the warming of the ambient temperature in the place between the first normalized record at 20 °C and that of the three following periods: from 20 to 24 °C, from 20 to 28 °C and from 20 to 32 °C.

The results of this evaluation and thermographic analysis were published.

6.2 Sensor Implementation Conclusion

This study introduces a novel approach for monitoring temperature responses in hyperhidrosis (HH) patients, utilizing a Python-based graphical application integrated with a Raspberry Pi and DHT22/AM2302 sensor. The system demonstrated its ability to reliably monitor skin temperature variations in response to thermal stimuli and successfully identified significant differences between HH patients and healthy controls. These findings contribute to the understanding of the underlying autonomic dysfunction associated with hyperhidrosis and present the system as a valuable tool for both clinical and research applications.

6.3 Key Findings

6. **Thermoregulatory Dysfunction in HH:** The significant differences observed in skin temperature between HH patients and controls, particularly in the palms and axillae, support the hypothesis that HH patients exhibit impaired thermoregulatory responses. Specifically, the reduced vasodilation and heat dissipation capacity in HH patients, compared to healthy controls, may provide valuable insights into the autonomic dysfunction characteristic of the condition.
7. **System Effectiveness:** The Raspberry Pi-based monitoring system proved effective in real-time data collection, providing visual and auditory alerts when temperature thresholds were exceeded. The integration of temperature and humidity sensors with real-time monitoring allows for continuous tracking, which is essential for assessing the condition and guiding therapeutic interventions.
8. **Clinical Relevance:** The study's findings underscore the potential clinical utility of the system. By monitoring autonomic responses in real-time, the system offers clinicians a non-invasive, objective tool for diagnosing HH, evaluating symptom severity, and adjusting treatment plans. This is especially important in a clinical environment where the management of hyperhidrosis is often based on subjective patient reports.

9. **Statistical Validation:** The statistical analysis, including t-tests and regression analysis, confirmed the presence of significant differences in temperature responses between the groups. This rigorous analysis lends credibility to the findings and strengthens the argument for the system's validity as a diagnostic tool.

6.4 Critical Evaluation of the Study and Limitations

While the study presents compelling evidence regarding the use of temperature monitoring as a diagnostic tool for hyperhidrosis, several important limitations must be considered:

10. **Controlled Environment:** The study was conducted under controlled laboratory conditions, which may not reflect the complexity of real-world environments. Factors such as fluctuating ambient temperature, humidity, and physical activity levels could influence the results, and these variables were not fully accounted for in the study. Future research should aim to replicate these findings in more naturalistic settings to validate the system's performance in everyday clinical practice.
11. **Sample Size and Generalizability:** The relatively small sample size of 37 HH patients and 16 controls limits the generalizability of the results. A larger and more diverse cohort would provide a more comprehensive understanding of how the system performs across different populations. Additionally, including a broader age range and varying degrees of HH severity would allow for a more nuanced analysis of the system's effectiveness and its potential in personalized treatment strategies.
12. **System Accuracy and Reliability:** While the system demonstrated its ability to monitor skin temperature effectively, there are areas for improvement in terms of sensor accuracy. The DHT22/AM2302 sensor, while suitable for the prototype, has limitations in terms of its precision and sensitivity at lower temperatures. Future versions of the system may need to incorporate more accurate sensors, such as thermistors or thermocouples, to ensure that the measurements remain reliable across a broader range of conditions.
13. **Long-Term Monitoring and Validation:** Although the system shows promise as a tool for short-term monitoring, further studies are needed to assess its reliability over extended periods. Continuous monitoring would allow for a better understanding of how thermoregulatory responses evolve over time, particularly in relation to treatment outcomes. Longitudinal studies could also help determine whether the system can be used to track disease progression or predict exacerbations in patients with HH.

6.5 Future Developments and Research Directions

The findings from this study open several exciting avenues for future research and technological development:

14. **Integration with Additional Sensors:** To enhance the comprehensiveness of the system, future versions could incorporate additional physiological sensors, such as those measuring skin perfusion, heart rate, or blood pressure. This would allow for a more complete assessment of autonomic dysfunction, which is central to the

pathophysiology of HH and other related disorders. Furthermore, integrating multi-sensor data could provide a more accurate picture of the patient's condition, enabling more precise diagnosis and treatment strategies.

15. **Remote Monitoring and Data Sharing:** One of the most promising developments for this system is the ability to integrate it into a remote monitoring platform. By enabling real-time data transmission to healthcare providers, the system could be used to continuously monitor patients without the need for constant clinical visits. This would be particularly valuable for patients who require long-term management or those in remote areas where access to healthcare is limited. Additionally, this could facilitate telemedicine applications, allowing for remote consultations and real-time adjustments to treatment plans.
16. **Machine Learning and Predictive Modelling:** To improve the system's diagnostic capabilities, machine learning algorithms could be implemented to analyse the collected data and identify patterns that might not be immediately apparent through traditional statistical analysis. For example, machine learning models could predict the likelihood of disease progression or therapeutic response based on historical temperature data, enabling more personalized treatment options. The use of predictive analytics could also help identify patients at risk of severe symptoms or complications, allowing for early interventions.
17. **Wearable Devices and Continuous Monitoring:** The next step in this research is to move beyond the laboratory setup and develop wearable devices that can continuously monitor temperature and other autonomic markers in HH patients. Wearable sensors would enable long-term tracking, providing continuous feedback on the patient's condition. This data could be crucial in monitoring the effects of treatment over time and in evaluating the long-term effectiveness of therapeutic interventions, such as iontophoresis or botulinum toxin injections.
18. **Multicentre Clinical Trials:** Finally, to fully assess the clinical utility of the system, multicentre clinical trials are essential. These trials would allow for a larger, more diverse sample of patients and help validate the system in real-world clinical settings. Furthermore, they would provide critical data on how the system integrates with existing clinical workflows and its impact on patient outcomes.

Acknowledgements. This research work was supported by National Funds through the Portuguese FCT-Fundação para a Ciência e a Tecnologia under the R&D Units Project Scope, UIDB/00760/2020 (https://doi.org/10.54499/UIDB/00760/2020).

References

1. Upton, E., Halfacree, G.: Raspberry Pi® User Guide. Wiley (2016). https://doi.org/10.1002/9781119415572
2. Ada, L: Adafruit DHT11, DHT22 and AM2302 Sensors. https://learn.adafruit.com/dht. Accessed 15 Feb 2025
3. Nuttall, B., Jones, D.: gpiozero - A simple interface to GPIO devices with Raspberry Pi. https://gpiozero.readthedocs.io/en/latest/. Accessed 15 Feb 2025

4. DiCola, T., Jaeschke, S., Fried, L. "Ladyada," Fang, K., Kastning, J.: GitHub - adafruit/Adafruit_Python_DHT: Adafruit Python DHT Sensor Library. https://github.com/adafruit/Adafruit_Python_DHT. Accessed 09 Feb 2025
5. PiGPIO, C.: pigpio library home page. https://abyz.me.uk/rpi/pigpio/examples.html. Accessed 15 Feb 2025
6. Contino, N., Scheller, A.: Raspberry Pi Hardware - Raspberry Pi Documentation

MultiFlow: An Ambient Intelligence Digital Twin

Diogo Torres[1], Eduardo Peixoto[1], Davide Carneiro[1,2](✉), Guilherme Palumbo[1,3], and Victor Alves[3]

[1] CIICESI, ESTG, Politécnico do Porto, 4610-156 Felgueiras, Portugal
{dpto,efrp,gfp}@estg.ipp.pt

[2] INESC TEC, R. Dr. Roberto Frias, 4200 Porto, Portugal
davide.r.carneiro@inesctec.pt

[3] ALGORITMI Research Centre/LASI, University of Minho, Braga, Portugal
valves@di.uminho.pt

Abstract. Ambient intelligence (AmI) refers to environments where smart devices, sensors, and AI-driven systems work seamlessly to enhance human interactions with their surroundings. Through the combination of real-time data, context-awareness, and adaptive learning, AmI enables environments to respond proactively to user needs, improving efficiency, comfort, and decision-making. However, since AmI systems are inherently human-centric and often operate autonomously, they must be designed with robust ethical, privacy, and safety considerations. Ensuring that these systems function reliably, fairly, and without harm is crucial, especially in sensitive domains like healthcare, security, and smart infrastructure. This work introduces a novel tool, conceptualized as an AmI Digital Twin, which allows developers to simulate or monitor AmI data streams, and develop and thoroughly test AmI applications before and during their real use. Built on a modular architecture leveraging technologies like React.js, Node.js, Kafka, Faust, MongoDB, InfluxDB, Grafana, and Docker, the platform ensures adaptability to different application environments, scalability, and ease of deployment. Besides the description of the tool itself, we provide some early validation results in common AmI tasks such as anomaly and concept drift detection. The tool is available in a public repository, and comes pre-packaged with a set of applications for AmI use-cases.

Keywords: Ambient Intelligence · Data Streams · Digital Twin

1 Introduction

In today's data-driven world, the increasing number of connected devices and sensors has transformed how we interact with our environments [6]. From smart homes to personal health monitoring systems, the ability to collect, process, and analyze real-time data has become essential for creating intelligent, adaptive, and responsive environments [4]. This ultimately led to the emergence of

M. Rodrigues et al. (Eds.): ISAmI 2025, LNNS 1776, pp. 12–19, 2026.
https://doi.org/10.1007/978-3-032-14138-5_2

Ambient Intelligence, where environments are aware of and adapt to the needs of individuals [2]. This evolution introduces two new main classes of challenges. The first derives from inherently human-centric and autonomous nature of AmI, which must ensure the well-being and safety of humans. The second are technical challenges, which derive from using complex and real-time streams of data to implement valuable, verifiable and testable human-centric applications.

Thus, it is paramount to appropriately test and validate AmI applications across multiple dimensions, including technical and ethical aspects, as these are fundamental to ensuring reliability and real-world readiness. Technically, these systems must be rigorously assessed for accuracy, robustness, and adaptability to dynamic environments. Ethically, they must be scrutinized for fairness, privacy protection, and bias mitigation to prevent unintended harm or discrimination. From a user-centric perspective, systems should be designed to integrate seamlessly into daily life while respecting autonomy and individual preferences [5]. Without comprehensive validation in these areas, AmI risks failing to deliver on its promise of enhancing human interactions safely and effectively.

To address these challenges, this work introduces MultiFlow[1]: a novel tool designed as a Digital Twin for Ambient Intelligence. Its main goal is, firstly, to simulate real-world data streams and enable rapid prototyping of streaming applications, enabling testing diverse use cases within ambient intelligence - such as anomaly detection or concept drift detection - and the assessment of algorithms or machine learning models under varying conditions. Then, at a later stage, it can be used to connect to a real AmI environment where these applications are deployed, provided typical technologies are used in that environment (e.g. Kafka). The tool thus facilitates the development and evaluation of applications, their transition to real-life deployment, and their continuous monitoring. The rest of the paper describes the application and its early validation.

This paper is structured as follows: Sect. 2 provides a background and related work on the key topics addressed. Section 3 details the design and implementation of MultiFlow. Section 4 presents the validation process and experimental setup, including the use of concept drift and outlier detection. Finally, Sect. 5 discusses the conclusions and potential future work.

2 Functional Scope

MultiFlow serves as an environment for experimenting with real data streams, both in real-time or sped-up, and prototyping streaming applications for AmI. By simulating realistic scenarios, developers can test different approaches, adjust their methods, and evaluate solution reliability in conditions that closely resemble real-world environments. This allows them to validate their work before deploying it in operational systems, such as smart homes, personal health monitoring, or connected devices. The platform offers several key features:

[1] https://github.com/davidecarneiro/multiflow.

- Project and Stream Management: Users can create projects to manage multiple related data streams within a single environment. Each stream is configured with a specific dataset and streaming parameters, such as the number of data points emitted per second or the total duration of the stream. These are operationalized through Kafka, which provides a modular and scalable backbone by assigning each data stream to a distinct topic. This design allows streams to operate independently and at different rates. Kafka's distributed architecture also ensures that high data loads can be handled efficiently, with the ability to scale horizontally by adding more brokers or partitions as needed. Streams simulate real-world ambient intelligence data sources, such as sensor data from smart homes or wearable devices.
- Custom Applications (*Apps*): MultiFlow *Apps* are Python-based scripts designed to handle specific streaming tasks, implemented in Faust. These can range from basic operations like data aggregation, filtering, or feature extraction, to more advanced algorithms such as detecting concept drift, identifying anomalies, or running machine learning models.
- Instances for Experimentation: Each *App* can spawn multiple *Instances*, which are variants of the base *App* with different configurations and input parameters. *Instances* allow users to test and compare various approaches or scenarios derived from the same foundational script, fostering rapid experimentation and iterative improvement. For example, developers can test an anomaly detection algorithm on different data streams to evaluate it under different scenarios. Alternatively, developers can also test different configurations of the same algorithm on the same stream, to find the optimum parameters.
- Visualization: Outputs from the *Instances* of each *App* are integrated with Grafana dashboards for visualization, enabling real-time monitoring and insights regarding the behavior of the tested approaches.

Figure 1 (left) depicts the *Apps* page, which shows the existing Faust applications designed to handle specific tasks within the intelligent environment. Each App defines the logic of the intended application, and also declares the required parameters for its operation. These parameters (e.g. configurations of a given algorithm that will be run) are provided by the user when they create *Instances* of *Apps*. This allows to configure the behavior of applications, and experiment different configurations. On the right side, Fig. 1 illustrates the Projects page, where developers can manage their projects and associated streams. This page includes features such as real-time progress bars with the real-time streaming status for easy monitoring of stream progress.

3 Use-Cases

MultiFlow is designed to address diverse challenges in human-centric streaming data environments and offers value across a wide range of use cases in Ambient Intelligence, some of which are described below. As the use-cases show, the flexibility of the tool enables developers to make data-driven decisions, test hypotheses, and refine approaches in controlled but realistic settings, making it a valuable

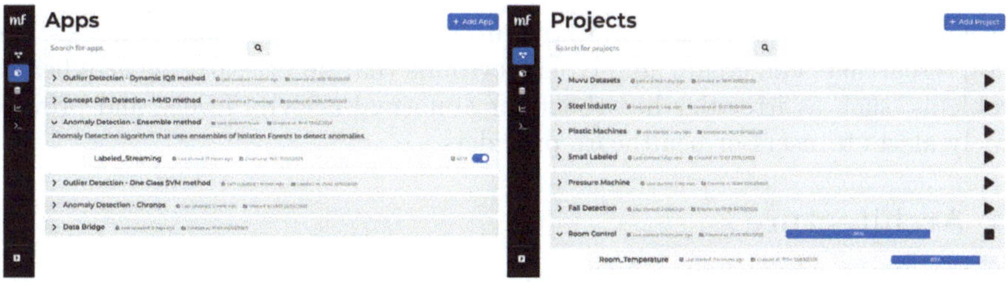

Fig. 1. Screenshots of the Projects page (right) showing stream management and progress visualization, and of the *Apps* page (left) listing the available applications and their *Instances*.

resource for advancing Ambient Intelligence applications, from smart homes to personalized health monitoring and beyond. While this tool was initially developed with the AmI domain in mind, its purpose can be extended to areas like Industrial IoT (e.g., predictive maintenance) and academic research (e.g., testing streaming ML algorithms).

Machine Learning Model Evaluation. Data scientists can use MultiFlow to simulate real-world conditions and test different machine learning models over time. By running multiple instances of the same application, they can compare how models behave under shifting data distributions or varying stream configurations [7]. This is particularly useful for applications such as predicting energy consumption in smart homes or personalizing health monitoring systems, enabling the identification of the most robust and effective models for deployment in dynamic environments.

Anomaly and Outlier Detection. Data analysts can use the tool to compare the efficacy of various anomaly detection algorithms. By simulating streams with anomalies or noisy data, they can assess how each method performs under realistic conditions [3]. Examples include detecting unusual patterns in home security systems, such as unexpected motion or door openings, or to identify irregularities in health metrics from wearable devices.

Concept Drift Detection. Researchers can test and refine adaptive learning systems for drift detection methods using scenarios with known distribution shifts, ensuring their solutions remain effective in dynamic environments, such as smart homes where user behavior or environmental conditions may change over time [7]. Examples include the gradual changes in activity patterns of aging people, or in people with onset of cognitive impairments.

Personalized Health Monitoring. MultiFlow can be used to simulate data streams from wearable devices, such as heart rate monitors or fitness trackers, to develop and test personalized health monitoring systems. By analyzing real-time data, researchers can create adaptive algorithms that detect early signs of health problems, such as irregular heartbeats or abnormal activity levels, and provide timely alerts to users or healthcare providers.

4 Architecture

The architecture of the tool is based on technologies commonly used in the implementation of intelligent environments, allowing an easy integration with real streams, and facilitating the transition from development to deployment.

The database layer uses MongoDB to manage project configurations, stream definitions, and metadata. The backend, built on Node.js, handles API services, project and stream creation, and integrates with external systems, while WebSocket ensures real-time communication.

The frontend, developed with React.js, offers an intuitive interface for configuring projects, defining streams, and managing the *Apps*. The streaming infrastructure relies on Kafka for high-throughput messaging and Faust for real-time stream processing. The *Apps* to be run are implemented in Faust, enabling the use-cases described above.

The visualization layer relies on InfluxDB for storing the output data from faust applications, and Grafana for interactive dashboards and real-time insights. The system's modular design allows components like the messaging system or database to be swapped or updated with minimal reconfiguration, ensuring adaptability to various scenarios of application.

In general, this architecture supports a variety of use cases, including the continuous evaluation of machine learning models under dynamic conditions, comparative analysis of anomaly detection systems [3], and experimentation with concept drift detection approaches [7]. It is ideal for applications like smart home automation, personalized health monitoring, and environmental sensing.

Figure 2 illustrates the architecture of MultiFlow, showing how its components interact and contribute to the overall functionality.

One relevant requirement of AmI architectures and applications, given that they deal with humans and human-derived data, is that they can be tested for ethical considerations, including fairness, bias or human control and oversight. This architecture was designed following principles of observability, in order to natively support the monitoring of ethical AI metrics throughout the whole pipeline. This is achieved through a series of AI ethics metrics, implemented as data processing pipelines with connectors by combining technologies for data integration from multiple sources, LLM-driven analysis, and real-time analytics to streamline the entire conformity assessment and oversight process. Specifically, we consider a range of previously identified metrics [5], and developed applications that monitor them across the pipeline. These metrics are domain-agnostic, and are calculated through data collected from the different open-source tools used in the architecture.

Some metrics are calculated based on the observation of operational data and employing message aggregation and sliding windows. Data freshness and quality,

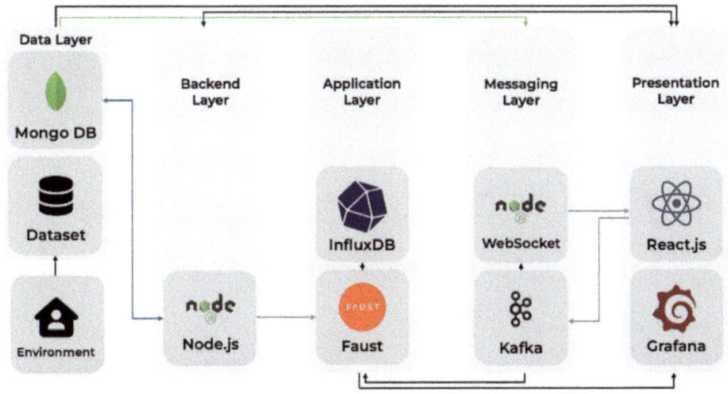

Fig. 2. System architecture highlighting data flow and component interactions.

for example, are calculated based on the data arriving through Kafka topics or previously stored data. Data imbalance is quantified by monitoring the data in databases, while bias (through demographic parity) is calculated through the difference between the probability of a prediction being positive between two different groups. Other metrics are process-related and are derived from the observation and monitoring of the code-base of the organization. These analyses are automated and based on compliance checks through LLMs. By following this observability-based approach, the tool facilitates the adhesion, monitoring and compliance with ethics standards.

5 Preliminary Validation

To assess the initial functionality and potential of the proposed tool, early validation was performed using key features that have already been implemented, showcasing its capabilities in real-world scenarios by simulating data streams and executing targeted applications. Two primary use cases spanning anomaly detection and concept drift detection, were selected for this validation due to their critical relevance in Ambient Intelligence systems.

Anomaly detection and concept drift detection were validated using datasets simulating falls and changes in data distribution, respectively, in a smart home environment for elderly care. A specifically-developed Faust App analyzed the sensor data, flagging anomalous events (e.g., falls of people) in real time and identifying shifts in data distribution [3,7], such as changes in a given room temperature (measured in Celsius), indicative of concept drift [1]. Results were visualized using Grafana, as shown in Fig. 3, which highlights deviations from expected behavior (anomaly detection) and emphasizes the application's ability to adapt to evolving conditions (concept drift detection). This dual functionality is particularly relevant in ambient intelligence scenarios, such as adapting to seasonal changes in user behavior in smart homes or evolving health metrics in wearable devices [1].

Fig. 3. Grafana dashboards showing fall detection through anomaly detection (top) and concept drift detection in room temperature (bottom).

Early validation confirms the tool's potential to simulate and analyze diverse Ambient Intelligence scenarios, supporting prototyping and testing of data-driven applications to reduce risks and improve decision-making in real-world deployments [3]. While the anomaly and concept drift detection use cases illustrate its value, the tool itself does not rely on any specific algorithms. Instead, it offers a flexible environment where developers can test and compare different approaches based on their own needs.

6 Conclusions and Future Work

This work introduced a novel tool for simulating data streams and prototyping data-driven applications in realistic Ambient Intelligence scenarios. By leveraging the concept of digital twins for data streams, the platform acts as a virtual replica of real-world environments, enabling the testing and optimization of machine learning applications in controlled yet realistic settings, and its later deployment and monitoring in real environments. Its modular architecture integrates technologies like Kafka, Faust, MongoDB, Grafana, and InfluxDB, ensuring scalability, adaptability, and ease of deployment.

The development of this application enabled the creation of pipelines for configuring and managing data streams, integrating custom applications, creating instances for comparative analysis, and visualizing results in real time. Validation through data streaming simulation tests, including use cases with anomaly detection and concept drift detection, demonstrated the tool's ability to effectively address these challenges. The positive results in these scenarios highlight the tool's potential as a robust environment for testing, experimentation, and decision support in data-intensive Ambient Intelligence applications, such as smart homes, personalized health monitoring, and environmental sensing.

Despite these achievements, opportunities remain to enhance and expand MultiFlow. Future work will focus on improving the tool's overall performance, including scaling to handle larger and more complex data streams while maintaining low latency and expanding support for advanced streaming parameters and customization options. These improvements, alongside the development of additional pre-configured applications, aim to position this tool as a comprehensive solution for real-time data analysis and decision support in diverse contexts, including Ambient Intelligence.

In conclusion, this tool simplifies the prototyping, testing, and validation of data-driven applications in realistic conditions, bridging the gap between research and practical deployment. By addressing future improvements, it has the potential to become a vital resource for innovation and decision-making in Ambient Intelligence scenarios.

Acknowledgements. This work has been supported by FCT – Fundação para a Ciência e Tecnologia within the R&D Units Project Scope: UID/00319/2023. The work of Guilherme Palumbo has been supported by national funds through FCT – Fundação para a Ciência e Tecnologia through project UIDP/04728/2020.

References

1. Guo, H., Li, H., Ren, Q., Wang, W.: Concept drift type identification based on multi-sliding windows. Inf. Sci. **585**, 1–23 (2022)
2. Martinez-Martin, N., et al.: Ethical issues in using ambient intelligence in healthcare settings. Lancet Digit. Health **3**(2), e115–e123 (2021)
3. Mohammadi, M., Al-Fuqaha, A., Sorour, S., Guizani, M.: Deep learning for IoT big data and streaming analytics: a survey. IEEE Commun. Surv. Tutorials **20**(4), 2923–2960 (2018)
4. Oliveira, J.D., Couto, J.C., Paixão-Cortes, V.S.M., Bordini, R.H.: Improving the design of ambient intelligence systems: guidelines based on a systematic review. Int. J. Hum.-Comput. Interact. **38**(1), 19–27 (2022)
5. Palumbo, G., Carneiro, D., Alves, V.: Objective metrics for ethical AI: a systematic literature review. Int. J. Data Sci. Anal. (2024). https://doi.org/10.1007/s41060-024-00541-w
6. Shah, S.H., Yaqoob, I.: A survey: internet of things (IoT) technologies, applications and challenges. In: 2016 IEEE Smart Energy Grid Engineering (SEGE), pp. 381–385 (2016)
7. Suárez-Cetrulo, A.L., Quintana, D., Cervantes, A.: A survey on machine learning for recurring concept drifting data streams. Expert Syst. Appl. **213**, 118934 (2023)

SallyEval: A Scalable Evaluation System for Ranking LLMs with Qualitative Feedback

Christos Kouros[1], Stergios Tegos[1], Alexandros Tassios[1], Konstantinos Manousaridis[1], Christos Bouas[1], Maria Papoutsoglou[1], Maria Kaltsa[2,3], Thanassis Mavropoulos[3], Stefanos Vrochidis[3], and Georgios Meditskos[1](✉)

[1] School of Informatics, Aristotle University of Thessaloniki, Thessaloniki, Greece
{ckouros,stegos,tassiosa,mmchristos,mpapo,gmeditsk}@csd.auth.gr

[2] Department of Theoretical and Applied Linguistics, School of English, Aristotle University of Thessaloniki, Thessaloniki, Greece
mkaltsa@enl.auth.gr

[3] Information Technologies Institute, Centre for Research and Technology Hellas, Thessaloniki, Greece
{mavrathan,stefanos}@iti.gr

Abstract. Large Language Models (LLMs) are widely used across various applications, yet their performance varies significantly depending on the task. Some models excel in reasoning, while others perform better in language understanding or factual accuracy, making it challenging to determine the most suitable LLM for specific use cases. To address this, we introduce SallyEval, an evaluation platform designed to identify the optimal LLM for integration into the Sally chatbot, an AI system that assists migrants in Greece. SallyEval employs A/B testing with human preference-based scoring, incorporating Elo rating and adaptive sampling to efficiently rank models based on real-world user interactions. In evaluations, the top-performing model, GPT4o-Latest, achieved an Elo score of 1050, with a 64% win rate. Unlike traditional pairwise comparisons that solely determine a winning response, SallyEval enhances the evaluation process by allowing evaluators to provide detailed feedback through a rating scale that covers multiple qualitative assessment criteria, such as accuracy, contextual relevance, clarity, and conciseness, while also supporting custom benchmarking for domain-specific evaluations.

1 Introduction

AI-powered chatbots are becoming increasingly common, helping people find information and receive support in various fields. Large Language Models (LLMs) like GPT-4o, Claude and Gemini have made chatbots more capable than ever, allowing for natural and informative conversations. However, the effectiveness of a chatbot depends heavily on the specific task, domain and language

M. Rodrigues et al. (Eds.): ISAmI 2025, LNNS 1776, pp. 20–30, 2026.
https://doi.org/10.1007/978-3-032-14138-5_3

requirements. While some models excel in reasoning and factual accuracy, others perform better in handling multilingual queries. This variation is particularly critical in applications where language plays a major role in user experience.

One such application is migration support, where AI chatbots assist individuals in navigating complex legal procedures, social services, government regulations, and more. These systems must provide clear and accurate information to users with diverse backgrounds and needs. Selecting the most suitable LLM for this domain is challenging, as existing benchmarks are often too general or rely on static datasets that overlook real-world user preferences [13].

Despite rapid progress in LLM development, the evaluation of these models remains an open research problem. Traditional benchmarks are based on predefined data sets with static questions and ground truth answers, such as MMLU [10], GSM-8K [4], and HellaSwag [11]. While these benchmarks provide useful quantitative insights, they fail to assess the effectiveness of LLMs in dynamic, real-world chatbot interactions, where user preferences play a crucial role. Furthermore, benchmarks that focus on factual accuracy do not account for conversational aspects, such as clarity and contextual adaptation, which are essential for user satisfaction. Beyond these challenges, the rapid evolution of LLMs further complicates the evaluation process, with new models being released frequently. Keeping evaluation frameworks up to date requires constant adaptation, and evaluating LLMs is often slow and resource-intensive.

To address these challenges, we introduce **SallyEval**, a systematic evaluation pipeline designed to compare and rank LLMs for integration into **SALLY**, a chatbot developed to assist migrants in Greece. The primary goal of this work is to identify the most suitable LLM to power SALLY by evaluating multiple models based on their ability to provide accurate, clear, and contextually relevant responses to migration-related queries. SallyEval's evaluation pipeline can be used by AI researchers, developers, data scientists, and organizations seeking to assess LLMs for diverse real-world applications. Unlike static benchmarks, SallyEval employs a pairwise human preference evaluation framework, leveraging A/B testing, Elo rating, and adaptive sampling to efficiently rank LLMs based on real interactions. While our approach builds upon recent advancements in LLM evaluation methodologies, such as Chatbot Arena's pairwise human preference ranking system [3], SallyEval extends this framework to address key limitations by focusing on **domain-specific assessment criteria** and qualitative feedback mechanisms, which are essential for evaluating the underlying LLMs used in task-specific applications like SALLY. In contrast, SallyEval allows for custom qualitative evaluation categories and predefined question sets, making it particularly suitable for applications such as migration support.

The contributions of this work can be summarized as follows:

- We propose a systematic evaluation pipeline for **LLM comparison**, integrating A/B testing, Elo ranking, and adaptive sampling to identify the best performing model for SALLY.
- We develop a dynamic evaluation framework, allowing users to compare responses from preselected LLMs, upload custom evaluation configurations

with domain-specific assessment criteria, and incorporate existing question-answer datasets.

- We demonstrate the use of SallyEval in evaluating the underlying LLMs for their suitability within the SALLY platform, identifying key factors that influence LLM effectiveness in this domain.

2 Related Work

Traditional evaluation methods for conversational systems fall into three main categories: automatic, human, and hybrid approaches

Automatic evaluation methods rely on predefined metrics to assess the performance of language models. Popular metrics such as BLEU, ROUGE and METEOR [5,6] compare generated responses against reference texts based on n-gram similarity. However, these methods often overlook important aspects like conversation flow, coherence, and contextual relevance, making them less effective for chatbot assessment. More recent approaches incorporate embedding-based similarity measures, such as BERTScore [12] and cosine similarity in vector spaces, providing more context-aware evaluations.

Human evaluation remains the most reliable method for assessing language models, offering qualitative insights into fluency, logical consistency, and user satisfaction. Common techniques include Likert-scale ratings, pairwise A/B testing [7], and user surveys. **A/B testing** is a comparative evaluation approach where users are presented with two responses to the same query, each generated by a different model, without knowing their source. Evaluators then select the response they find superior. Studies indicate that human evaluators prioritize attributes such as clarity, informativeness, and engagement over lexical similarity to reference responses [1].

Hybrid approaches aim to balance scalability and evaluation reliability by integrating automated metrics with human preference judgments. One such approach is the **Elo rating system**, a ranking method originally designed for competitive games, where models are dynamically scored based on head-to-head comparisons. When an evaluator selects a preferred model response, the winning model's Elo score increases while the losing model's score decreases [2]. To improve ranking efficiency, many evaluation systems utilize **adaptive sampling**, a technique that dynamically selects which models to compare based on their evaluation confidence. Instead of choosing models randomly, adaptive sampling prioritizes comparisons involving under-tested or closely ranked models. For instance, suppose SallyEval is evaluating three LLMs: Model A, Model B, and Model C. Initially, all models have similar Elo scores due to limited evaluation data. As feedback is collected, Model A consistently outperforms others, while Model C underperforms. Model B, however, has fewer evaluations, making its ranking uncertain. Instead of continuing random pairwise comparisons, SallyEval adaptively selects more comparisons involving Model B against both A and C. This targeted approach reduces uncertainty in ranking, ensuring a more efficient allocation of resources, particularly in large-scale model comparisons [8].

While these evaluation methods are effective for general language model assessment used in chatbots, they often fail to account for the challenges of domain-specific applications, such as legal advisory, healthcare, migration support and others. As a result, there is a growing need for tailored benchmarking approaches that consider task-specific requirements. Several key challenges remain:

- **Lack of task-specific metrics**: Most existing benchmarks, such as MMLU, GSM8K, and HellaSwag, primarily assess coding, mathematics, or general knowledge, rather than accuracy in specialized fields like legal consultation, administrative processes, and migration support.
- **Bias in human evaluation**: User feedback is influenced by personal expectations, cultural factors, and prior chatbot interactions, leading to inconsistencies in assessment outcomes.
- **Scalability limitations**: Large-scale human evaluations require significant time and resources, highlighting the need for hybrid methodologies that combine automated scoring with user-driven feedback.

To bridge these gaps, SallyEval introduces a structured, scalable evaluation process. Unlike traditional benchmarks, it allows for flexible, user-defined settings, making it especially useful for critical applications while offering valuable insights.

3 Methodology

SallyEval evaluates LLMs using a structured benchmarking pipeline that ensures consistency, reproducibility and adaptability (Fig. 1).

- **Step 1: LLM Selection (Hidden from User).** The system automatically selects two LLMs for evaluation. This selection follows an adaptive sampling strategy (see Sect. 2) that prioritizes comparisons involving models with uncertain rankings or limited evaluation data, ensuring a more accurate ranking over time.
- **Step 2: User Poses a Question.** Once the models are selected, the user initiates the conversation by submitting a query. Each LLM independently generates a response within its own chat session, allowing the user to engage with both models separately.
- **Step 3: Dynamic Chat Interaction.** The user can continue the conversation with both models, exchanging messages iteratively. Since the chat histories develop independently for each model, their responses may differ in tone, coherence, and contextual adaptation, reflecting the variations in their conversational abilities.
- **Step 4: Response Evaluation.** At any point during the conversation, the user can evaluate the responses by selecting the one they perceive as superior. To provide further insights, the user can also assign a numerical rating from 1 to 5 for each evaluation category.

- **Step 5: LLM Identities Revealed.** After submitting their evaluation, the system reveals the identity of each LLM, allowing users to see which model generated which response. This approach prevents bias during the assessment phase while enabling post-evaluation review of model performance.
- **Step 6: Iterative Process.** Users may initiate a new question and repeat the evaluation process as needed, ensuring continuous assessment and refinement of model rankings based on real user interactions.

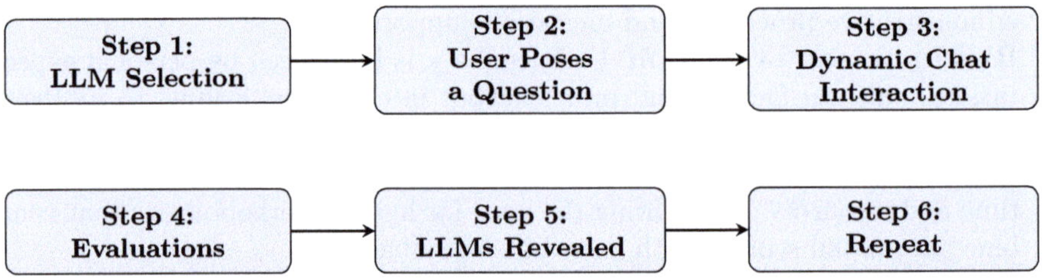

Fig. 1. Flowchart of the SallyEval Evaluation Pipeline for A/B Testing

3.1 Custom Evaluation Configurations

While SallyEval allows for predefined evaluation categories such as accuracy, clarity, and contextual relevance, these serve as qualitative assessment criteria rather than static benchmarks. However, users can define domain-specific benchmarks by uploading question datasets, which will be answered by all LLMs in the system, enabling a more structured evaluation process. Additionally, users can set custom evaluation categories for each benchmark, allowing for tailored assessments based on specific domain requirements.

- **Custom benchmark categories:** Upload custom question datasets to create structured evaluation scenarios for all LLMs.
- **Custom evaluation categories:** Define specific rating criteria that can vary based on the selected question dataset (benchmark).
- **Manual LLM selection:** Compare specific LLM pairs using a **blind evaluation process** (model identities remain hidden until submission).

3.2 Evaluation Metrics

Evaluations capture both quantitative and qualitative metrics, including response time and detected language. User assessments provide additional insights through pairwise evaluations, where evaluators compare responses and select the preferred one. Along with their selection, users also provide a numerical rating (1 to 5) for each evaluation category, offering a structured justification for their choice of LLM.

SallyEval tracks comparison counts and updates model rankings using the Elo rating system, applying larger score adjustments for unexpected wins. Each evaluation is timestamped, allowing for the analysis of performance trends over time.

The Elo rating system updates both the winner's and loser's scores after each comparison using the following formulas:

$$E_{\text{winner,new}} = E_{\text{winner,old}} + K \times (1 - P_{\text{winner}}) \tag{1}$$

$$E_{\text{loser,new}} = E_{\text{loser,old}} + K \times (0 - P_{\text{loser}}) \tag{2}$$

where:

- $E_{\text{winner,new}}$ and $E_{\text{loser,new}}$ are the updated Elo scores for the winning and losing models.
- $E_{\text{winner,old}}$ and $E_{\text{loser,old}}$ are the previous Elo scores.
- K is the scaling factor that controls the sensitivity of rating adjustments.
- P_{winner} and P_{loser} are the expected probabilities of winning for each model, computed as:

$$P = \frac{1}{1 + 10^{(E_{\text{opponent}} - E_{\text{current}})/400}} \tag{3}$$

This ensures that models with lower initial ratings gain more Elo points when winning against higher-ranked models, leading to a more dynamic and adaptive ranking system. The structured collection of evaluation data enables a deeper understanding of user preferences and model effectiveness, supporting ongoing improvements to benchmarking methodologies.

3.3 Implementation Details

SallyEval is implemented as a web-based evaluation platform using a client-server architecture. The frontend is built with React, selected for its component-based architecture as it promotes reusability and maintainability, making it easier to scale the platform as new features are introduced.

The backend is built with Flask, a Python web framework chosen for its simplicity, extensive libraries, and strong community support. Python's flexibility enables efficient handling of API requests. The backend contains RESTful API endpoints, with WebSocket integration enabling real-time updates where necessary. SallyEval uses MongoDB Atlas to store and manage all collections efficiently. The database is optimized for fast read and write operations, with indexed fields improving query speed and ensuring the system remains scalable as evaluation data grows.

4 Use Case: Migration Support

SALLY [9] is an AI chatbot designed to assist immigrants in Greece. The chatbot is built on a RAG architecture to combine information retrieval with generative AI, ensuring accurate, context-aware responses. This approach ensures that answers are grounded in factual sources while minimizing the risk of misinformation. SALLY primarily sources information from migration.gov.gr, the official website for migration policies in Greece. By continuously updating its knowledge base, SALLY ensures that its responses remain accurate, relevant, and aligned with official regulations.

However, evaluating the performance of the underlying LLMs used in SALLY presents unique challenges that traditional LLM evaluation frameworks do not fully address. As previously mentioned, existing benchmarks often assess general-purpose reasoning, factual correctness, or coding abilities, but they fail to capture domain-specific factors which are critical for applications such as assisting migrants. A challenge that SallyEval is built to address.

4.1 Dataset

The dataset was built using real-world scenarios that migrants often encounter, ensuring the content was both relevant and actionable. The benchmarks address two critical aspects of migrant support:

Legal and Immigration Procedures. This benchmark assesses the ability of LLMs to deliver accurate, legally sound, and actionable advice on immigration laws and procedures in Greece. The dataset was curated using questions from legal forums, government sources, and real-world queries from migrants. For example, a common question might be, *"How do I apply for asylum in Greece?"* Given the complexity of legal processes, ensuring accurate and up-to-date responses is crucial. Misinterpretations can lead to severe consequences, such as deportation or invalid applications. Additionally, the dynamic nature of immigration regulations requires LLMs to provide reliable guidance despite frequent legislative changes. Furthermore, the procedural steps involved in immigration are often complex and bureaucratic, necessitating clear, step-by-step instructions to avoid confusion and errors.

Practical Assistance and Social Integration. This benchmark evaluates the LLM's ability to provide practical, culturally relevant and accessible support to migrants adapting to life in Greece. The dataset includes typical questions about daily life, such as housing, healthcare, education and social services. For instance, a common query might be, *"How do I register my child for school in Greece?"*. In addition to legal guidance, this benchmark emphasizes the importance of addressing everyday challenges. Migrants often face unique cultural and linguistic barriers, requiring clear and culturally aware responses. Effective support in this area can significantly impact a migrant's ability to navigate their new environment and integrate into the local community.

4.2 Data Collection

The dataset consisted of a total of 100 chats, which included not only initial questions but also follow-up questions when necessary to ensure a comprehensive evaluation. All nine authors, each with a technical background, participated in the evaluation process, contributing a diverse range of perspectives and expertise. Evaluators also impersonated migrants to better assess SALLY's real-world effectiveness.

While the number of LLMs used in the evaluation was relatively small (three in total), the aim was to present the potential of this evaluation framework. In future versions, more LLM options will be incorporated to expand the evaluation scope and further refine the system's capabilities.

Through these evaluation tools, we were able to measure and compare not only the LLM's performance but also the chatbot's user satisfaction and real-world effectiveness in assisting migrants.

Table 1. Performance Metrics of Models

Model	ELO	Win %	Lose %	Rating
4o-Latest	1050.63	64	36	4.4
Flash-2o	960.83	44	56	4.19
Claude 3.7 Sonnet	988.54	47	53	4.0

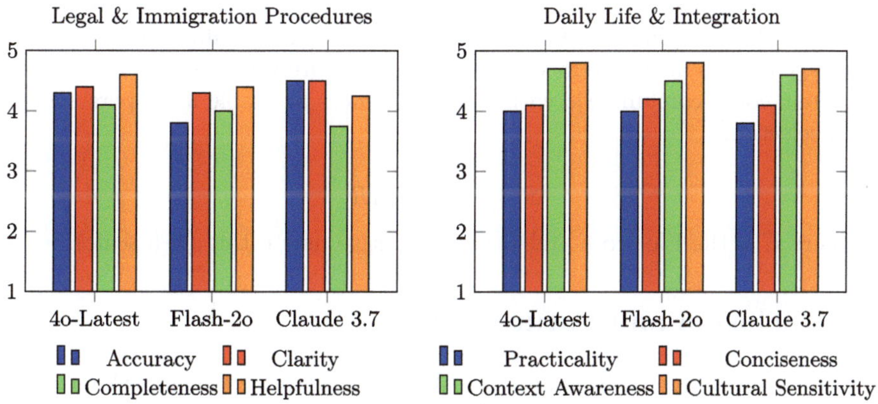

Fig. 2. Comparison of Evaluation Scores.

4.3 Results

The data (Table 1 and Fig. 2) indicate that **GPT4o-Latest** outperformed the other models across most metrics, achieving the highest **Elo score of 1050.63** and a **win rate of 64%**, demonstrating its overall superiority in human preference-based evaluations. Notably, it received the highest **helpfulness rating (4.6)**, suggesting that users found its responses particularly useful in guiding them through legal and practical procedures. **Claude 3.7 Sonnet** performed competitively, obtaining the highest **accuracy (4.5)** and **clarity (4.5)** scores, indicating its strength in delivering precise and well-structured information. However, its **lower completeness score (3.75)** suggests that its responses may sometimes lack detail or fail to provide comprehensive answers.

The performance metrics in Table 1 include the Elo score, calculated using the method described in Sect. 3.2, where each comparison updates the score based on the expected win probability. Win and Lose percentages represent the share of pairwise comparisons each model won or lost, respectively. The Rating is the average score assigned by evaluators across the qualitative categories reflecting overall user satisfaction.

Meanwhile, **Flash-2o** showed a balanced performance with a clarity rating of **4.3** and a helpfulness score of **4.4**, but its **lower accuracy (3.8)** and **Elo rating (960.83)** indicate that it was generally less preferred compared to the other models. Overall, based on our benchmark results and SallyEval's current Elo-based ranking system, **4o-Latest** would be selected as the top-performing model, demonstrating strong user satisfaction across both legal and practical queries. **Claude 3.7 Sonnet** remains a strong contender in accuracy and clarity but may require improvements in completeness.

4.4 LLM Selection

While SallyEval ranks LLMs based on Elo scores, final model selection could also take qualitative insights into account. Users may examine evaluation results such as average ratings in the feedback categories. In future versions, a weighted scoring system could combine Elo with user preferences to enable more personalized model selection.

$$S = 0.6E + 0.4R \tag{4}$$

where E is the Elo score and R is the average user rating (1–5 scale).

5 Conclusion and Future Work

This study introduces SallyEval, a scalable framework for benchmarking LLMs using structured A/B testing and data-driven assessments. Unlike traditional methods, SallyEval supports customized evaluations, making it ideal for domain-specific use cases In our evaluation, the top-performing model, GPT4o-Latest, achieved an Elo score of 1050.63, with a 64% win rate.

However, the framework also comes with certain limitations. For example, evaluator subjectivity and potential biases can influence results, especially when user feedback is inconsistent. Additionally, maintaining up-to-date rankings in a rapidly evolving LLM landscape presents ongoing challenges, especially when new models are introduced frequently.

Future work will focus on addressing these limitations and expanding the framework's capabilities. Key priorities include integrating bias detection mechanisms to improve fairness, automating re-evaluation strategies to keep rankings current, expanding multilingual evaluation support to ensure fair assessments across languages, and integrating more LLMs to provide up-to-date performance comparisons. Furthermore, we aim to introduce more dynamic benchmarking configurations, such as weighting evaluation criteria based on task-specific priorities, enabling more targeted and context-aware assessments.

By tackling these challenges, SallyEval will continue to evolve into a more robust, fair, and insightful evaluation tool for large language models.

Acknowledgments. This research project is carried out within the framework of the National Recovery and Resilience Plan Greece 2.0, funded by the European Union - NextGenerationEU (Implementation body: HFRI, Project Number: 15010).

References

1. Adiwardana, D., et al.: Towards a human-like open-domain chatbot. arXiv preprint arXiv:2001.09977 (2020)
2. Boubdir, M., Kim, E., Ermis, B., Hooker, S., Fadaee, M.: Elo uncovered: robustness and best practices in language model evaluation. arXiv preprint arXiv:2311.17295 (2023)
3. Chiang, W.L., et al.: Chatbot arena: an open platform for evaluating LLMs by human preference. In: Proceedings of the 41st International Conference on Machine Learning (ICML 2024) (2024)
4. Cobbe, K., et al.: Training verifiers to solve math word problems. arXiv preprint arXiv:2110.14168 (2021)
5. Davoodijam, E., Alambardar Meybodi, M.: Evaluation metrics on text summarization: comprehensive survey. Knowl. Inf. Syst. **66**(12), 7717–7738 (2024)
6. Lins, R.D., Oliveira, H., Simske, S.J.: Assessing the reliability and validity of the measures for automatic text summarization. In: Proceedings of the ACM Symposium on Document Engineering 2024. Association for Computing Machinery (2024)
7. Liu, Y., Zhou, H., Guo, Z., Shareghi, E., Vulić, I., Collier, N.: Aligning with human judgement: the role of pairwise preference in large language model evaluators. arXiv preprint arXiv:2403.16950 (2024)
8. Lookman, T., Balachandran, P.V., Xue, D., Yuan, R.: Active learning in materials science with emphasis on adaptive sampling using uncertainties for targeted design. npj Comput. Mater. **5**(1), 1–11 (2019)
9. Meditskos, G., et al.: Towards semantically conscious, conversation-based chatbot services for migrants. In: Artificial Intelligence Applications and Innovations, pp. 139–148. Springer Nature Switzerland, Cham (2024)
10. Wang, Y., et al.: MMLU-Pro: a more robust and challenging multi-task language understanding benchmark. In: Proceedings of the 38th Annual Conference on Neural Information Processing Systems (2024)
11. Zellers, R., et al.: HellaSwag: can a machine really finish your sentence? In: 57th Annual Meeting of the Association for Computational Linguistics, pp. 4791–4800 (2019)

12. Zhang, T., Kishore, V., Wu, F., Weinberger, K.Q., Artzi, Y.: BERTscore: evaluating text generation with BERT. In: Proceedings of the 8th International Conference on Learning Representations (ICLR) (2020). https://openreview.net/forum?id=SkeHuCVFDr
13. Zheng, L., Chiang, W.L., Sheng, Y., et al.: Judging LLM-as-a-judge with mt-bench and chatbot arena. In: 37th International Conference on Neural Information Processing Systems (2023)

Bio-inspired Measurement Timing Coordination for Mobile Sensor Networks

Takashi Ikegami[1(✉)] and Ichiro Satoh[2]

[1] SOKENDAI/National Institute of Informatics, The Graduate University for Advanced Studies, Tokyo, Japan
takashi_ik@nii.ac.jp
[2] National Institute of Informatics, Tokyo, Japan
ichiro@nii.ac.jp

Abstract. Modern sensor networks often consist of larger numbers of mobile sensor nodes. In such sensor networks, multiple sensor nodes can measure common targets. If they measure their targets simultaneously, the impact of measurement errors can be reduced. Alternatively, if they perform their measurements at the same frequency while offsetting their measurement timings, temporal resolution can be enhanced. However, since a sensor network composed of mobile sensor nodes changes its topology dynamically, it becomes difficult to coordinate their measurement timing. In this work, we propose bio-inspired algorithms for the coordination of the measurement timing: synchronization, where both the periods and phases of sensor nodes are synchronized, and desynchronization, where their periods are synchronized but their phases are evenly offset. We evaluate our proposed synchronization method for these algorithms through a simulation.

Keywords: Sensor networks · Bio-inspired · Synchronization · Desynchronization

1 Introduction

Sensor networks have been used in a broad range of domains [3]. A sensor network consists of multiple sensor nodes, connected through wired or wireless networks. They periodically measure their targets. Conventional sensor networks were usually composed of as few sensor nodes as possible to avoid overlap in the measurement areas of sensor nodes. Sensor nodes rarely measure the common target. They are typically stationary once deployed. On the other hand, modern sensor networks can be composed of a larger number of sensor nodes. They can measure the common target at the same time. They can be mobile when they are attached to a mobile node, for example, an Unmanned Aerial Vehicle (UAV).

There are two advantages when multiple sensor nodes measure common targets. One advantage is the reduced impact of measurement errors. Measurement data typically contains some degree of error. The impact of these errors can be

M. Rodrigues et al. (Eds.): ISAmI 2025, LNNS 1776, pp. 31–40, 2026.
https://doi.org/10.1007/978-3-032-14138-5_4

reduced by comparing measurement data from multiple sensor nodes. Synchronizing the timing of their measurement of multiple sensor nodes that measure the common targets can enable this advantage. Another advantage is the ability to capture rapid changes in a measurement target that a single sensor node may not effectively capture. This advantage can be achieved by synchronizing the measurement periods of sensor nodes that measure the same targets, and by evenly offsetting their measurement timings to enhance temporal resolution.

We propose a method for the self-configured coordination of the measurement periods and phases, inspired by synchronization phenomena observed in biological systems. In nature, animals observe the rhythmic behaviors of others and coordinate their periodic behaviors. For example, fireflies flash in unison, and frogs try to avoid overlapping their calls. We model the periodic measurements of each sensor node in a sensor network as a set of phase-coupled oscillators. In this method, sensor nodes exchange phase information and coordinate their periods and phases based on the model. This method is designed to achieve either *synchronization* or *desynchronization*. *Synchronization* means that sensor nodes measure their targets simultaneously. On the other hand, *desynchronization* means that sensor nodes measure their targets at the same frequency while the phases of their measurement timings are evenly offset among sensor nodes. Users can choose either synchronization algorithms or desynchronization algorithms based on their purposes.

Synchronization algorithms have been proposed for sensor networks [9]. Desynchronization algorithms also have been proposed for sensor networks to coordinate the timings of wireless communication to avoid congestion in wireless communication [1,2,6,10]. Existing works have not specifically focused on coordinating the measurement timings of sensor nodes in sensor networks. Existing desynchronization algorithms that aim to avoid congestion in wireless communication assume that each node can directly communicate with all other sensor nodes. However, a node in a large-scale network may communicate with other nodes through more than one node. This assumption limits the scalability of these algorithms. Furthermore, such desynchronization algorithms only require ensuring that communication timings do not overlap to avoid wireless communication congestion. In contrast, maintaining a consistent measurement period may be critical when coordinating measurement timings. For these reasons, existing models are unsuitable for the self-configured coordination of measurement timings in sensor networks.

The paper is structured as follows: Sect. 2 describes the two proposed algorithms: the synchronization algorithm and the desynchronization algorithm. Section 3 provides the system design of a sensor network with them. Section 4 presents an evaluation of them through a simulation and discusses the evaluation results. Section 5 reviews related works. Finally, Sect. 6 concludes this paper, summarizing key points and future directions.

2 Proposed Methodologies

We describe the two proposed algorithms: synchronization and desynchronization for the coordination of measurement timings among multiple sensor nodes.

We outline our target sensor networks. Sensor nodes are connected through wired or wireless networks, and they can exchange messages with each other. There is no central computer or sensor node that manages the entire system in the sensor networks. We assume that any clock synchronization mechanisms (e.g., GPS-based synchronization protocols or Network Time Protocol (NTP)) are not available in the sensor networks. Sensor nodes can be dynamically added to or removed from the sensor networks during operation.

2.1 Synchronization

Synchronization means that multiple sensor nodes perform periodic measurements at the same time. Some animals in nature coordinate their periodic behaviors by observing others. An example is fireflies flashing in unison. We consider periodic measurements as phase oscillators. They then exchange their phases with each other to coordinate their measurement timings, as illustrated in Fig. 1.

An existing model demonstrated that synchronization can be facilitated by introducing a single pacemaker node in a network [8]. This model assumes that messages are received immediately after being sent by a node, and all nodes are connected with the pacemaker node directly. In practical sensor networks, sensor nodes exchange messages via more than one node, with inherent communication delays. This assumption limits the scalability of the model. We extend the model to take communication delays into account and introduce some hub nodes. Other sensor nodes try to synchronize with the phase of the hub nodes as described in Eq. (1), Eq. (2), and Algorithm 1.

$$\theta_h(t + \Delta t) - \theta_h(t) = \Delta t \cdot \left(\frac{1}{T_h} + \alpha \sum_{j=1}^{N} \Gamma\left(\theta_j(t - \tau_{hj}) - \theta_h(t)\right) \right) \quad (1)$$

$$\theta_i(t + \Delta t) - \theta_i(t) = \Delta t \cdot \left(\frac{1}{T_i} + \alpha \cdot \Gamma\left(\theta_h(t - \tau_{ih}) - \theta_i(t)\right) \right) \quad (2)$$

Here, $i = 1, 2, \ldots, N$ refers to the i-th sensor node and N denotes the number of nodes with which the i-th sensor node can communicate. θ_h denotes the phase of the hub node, and θ_i represents the phase of the i-th sensor node that is not a hub node. τ_{ij} denotes the communication delay between nodes. $\Gamma(\theta)$ denotes a periodic function with a period of 2π. In our proposed method, we use $\Gamma(\theta) = \sin\left(\theta_j(t - \tau_{hj}) - \theta_h(t)\right)$ Sensor nodes measure their targets when their phases θ_i (θ_h) reach 2π, and then reset their phases θ_i (θ_h) to 0. T_i (T_h) denotes the initial measurement period. α denotes a weight factor that determines the amount of change in a single step and affects the convergence speed. t denotes the local time of each sensor node and does not have to be synchronized.

Algorithm 1. Synchronization for each sensor node

Require: θ_i, θ_j (for all neighboring j-th nodes), ω_i, α, N, t_{current} (current local time), t_{prev} (previous measurement local time)

```
if this node is a hub node then
    sum ← 0
    for each neighboring j-th node do
        sum ← sum + 2π/ω + α · sin(θ_j − θ_i)
    end for
    v_θ ← sum/N
else if i-th node is not a hub node then
    if j-th node is a hub node then
        v_θ ← 2π/ω + α · sin(θ_j − θ_i)
    end if
end if
t_current ← getCurrentLocalTime()
Δt ← t_current − t_prev
θ_i ← θ_i + v_θ · Δt
if θ_i > 2π then
    Measurement()
    θ_i ← 0
end if
t_prev ← t_current
```

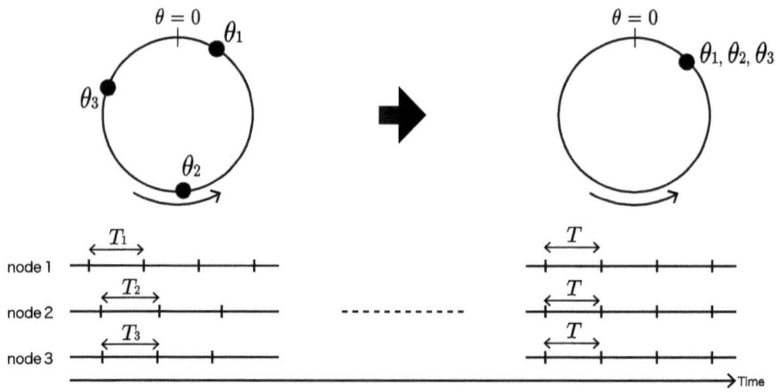

Fig. 1. Synchronization

We describe the procedure to determine hub nodes, as illustrated in Algorithm 2. Each sensor node has a randomly generated value $b_i \in [0, 1]$. Based on b_i, hub sensor nodes are determined by the following procedure: (1) If b_i of sensor nodes exceeds a preconfigured threshold, they are designated as hub

Algorithm 2. Determination of hub nodes

Require: b_i, b_j (for all neighboring nodes j), $b_{\text{threshold}}$

```
b_max ← 0
for each neighboring node j do
  if b_j > b_threshold then
    Record the ID of j-th node as a hub node
  else if b_j > b_max then
    Record the ID of j-th node as a hub node
    b_max ← b_j
  end if
end for
if b_i > b_threshold or b_i > b_max then
  Mark this node as a hub node
else
  Mark this node as not a hub node
end if
```

nodes. (2) If b_i of sensor nodes is the highest among the sensor nodes within their communication range, they are designated as hub nodes even if b_i of them is below the threshold.

Non-hub sensor nodes store the ID of the sensor node with the largest b_i among the sensor nodes with which they have communicated. They then try to synchronize the periods and the phases of their measurement timings as shown in Eq. (2).

2.2 Desynchronization

Desynchronization means that multiple sensor nodes perform periodic measurement at the same frequency among multiple sensor nodes while offsetting their phases, as shown in Fig. 2.

To achieve desynchronization, an offset δ_i is added to the synchronized phase. All sensor nodes maintain the same period, but the phases are offset evenly. The number of phase segments, D, into which the period is evenly divided, is assumed to be preconfigured. Increasing the value of D enables a sensor network to capture faster changes in a measurement target. The algorithm for calculating δ_i is shown in Algorithm 3.

The i-th node has a phase $\varphi_i(t)$ in addition to the phase θ_i of its measurement timing. The phase $\varphi_i(t)$ is adjusted to maximize its distance from the phases $\varphi_j(t)$ of neighboring nodes, as described in Eq. (3). In practice, information on the phase is transmitted with a communication delay τ_{ij}. The weight factor β

controls the step size of the coordination and influences the convergence speed.

$$\varphi_i(t + \Delta t) = \varphi_i(t) - \frac{\beta}{N}\sum_{j=1}^{N}\frac{1}{\varphi_j(t-\tau_{ij}) - \varphi_i(t)}, \qquad i = 1, 2, \ldots, N, \quad (3)$$

Since each sensor node may not be able to communicate with all others, it often lacks complete information about the entire network. To address this limitation, $\varphi_i(t)$ is used to distribute the sensor nodes as evenly as possible across equally divided intervals of the period. The offset δ_i is then assigned based on the interval containing $\varphi_i(t)$.

For example, when the number of segments into which the period is evenly divided D is 3, as illustrated in Fig. 3, the offset δ_i is determined as follows: if φ_i is included in $0 \le \varphi_i < \frac{2\pi}{3}$, then $\delta_i = 0$; if φ_i is included in $\frac{2\pi}{3} \le \varphi_i < \frac{4\pi}{3}$, then $\delta_i = \frac{2\pi}{3}$; and if φ_i is included in $\frac{2\pi}{3} \le \varphi_i \le 2\pi$, then $\delta_i = \frac{4\pi}{3}$. As a result, δ_1 and δ_2 is 0. δ_3 is $\frac{2\pi}{3}$. δ_4 and δ_5 is $\frac{4\pi}{3}$.

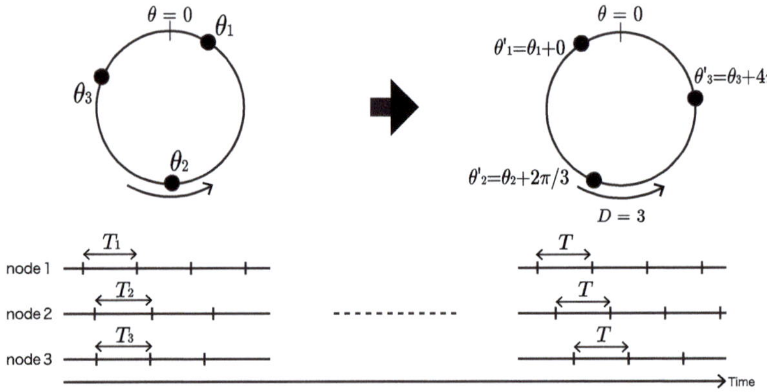

Fig. 2. Desynchronization

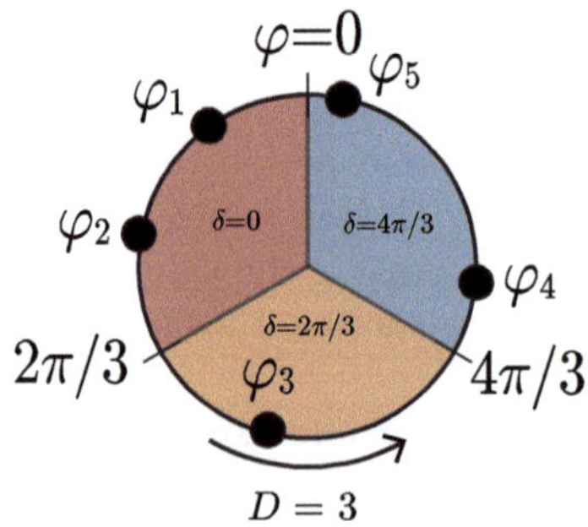

Fig. 3. An example of desynchronization

Algorithm 3. Desynchronization for each sensor node

Require: θ_i, φ_i, φ_j (for all neighboring j-th nodes), D
  $sum \leftarrow 0$
  for each neighboring sensor node j **do**
    $sum \leftarrow sum + \frac{1}{\varphi_j - \varphi_i}$
  end for
  $sum \leftarrow (sum/N) \cdot \beta$
  $\delta_i \leftarrow \lfloor \varphi_i \cdot \frac{D}{2\pi} \rfloor \cdot \frac{2\pi}{D}$
  if $\theta_i + \delta_i > 2\pi$ **then**
    Measurement()
    $\theta_i \leftarrow -\delta_i$
  end if

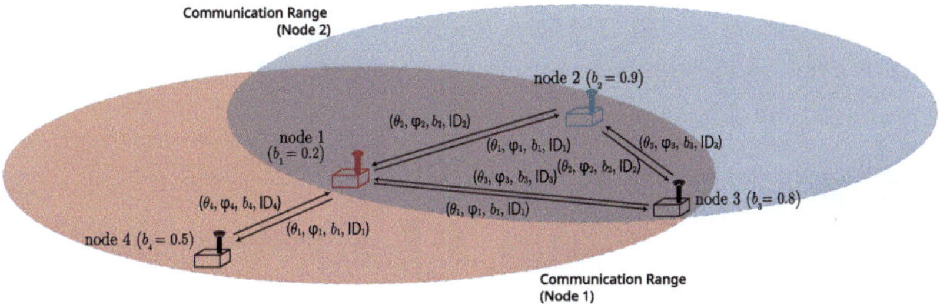

Fig. 4. An example of communication between sensor nodes

3 System Design

We describe the system design of a sensor network with the proposed algorithms. We assume that our target sensor network has no computer or sensor node that centrally manages the whole sensor network and no mechanisms to synchronize the clocks equipped in sensor nodes. Sensor nodes may be dynamically added to or removed from the sensor network. Each sensor node has a unique ID, the phase θ_i of its measurement timing, the phase φ_i, and a comparable value b_i. The phase θ_i of each sensor node does not necessarily start at 0. The phase φ_i used for desynchronization is randomly initialized within the range $[0, 2\pi]$. b_i is randomly initialized between 0 and 1. For desynchronization, the number of segments into which the period is evenly divided, D, is preconfigured. The larger the value of D, the better the sensor network can capture rapid changes in the target. Each sensor node multicasts its ID, its phase θ_i, its phase φ_i, and its b_i several times per cycle to sensor nodes within its communication range. An example of the sensor network is illustrated as shown in Fig. 4.

4 Evaluation

In this section, we evaluate our proposed synchronization and desynchronization algorithms. We develop a simulator implemented in Java to conduct a simulation. We use the parameters as follows: $\alpha = 1.0 \times 10^{-3}$ in Eq. (1), $\beta = 5.0 \times 10^{-5}$ in Eq. (3), and $D = 3$. In this simulation, 50 sensor nodes are randomly deployed within a circular area of radius 25 m. They move around within the circular area at a speed of 1 m/s. The communication range of each sensor node is 10 m. Under these conditions, a sensor node cannot exchange information directly with some of the sensor nodes in the sensor network. The initial measurement period is set to 1 s. Each sensor node multicasts information to all other nodes five times per cycle. If a sensor node does not receive new information from a specific node for more than 5 s, the node is treated as disconnected from the network. The simulation runs for 60 s.

Figure 5 illustrates simulation results of the proposed synchronization algorithm in a sensor network with 50 mobile sensor nodes. Figure 6 illustrates simulation results of the proposed desynchronization algorithm in a sensor network with 50 mobile sensor nodes. The synchronized or desynchronized state is stably maintained even as the network topology changes dynamically.

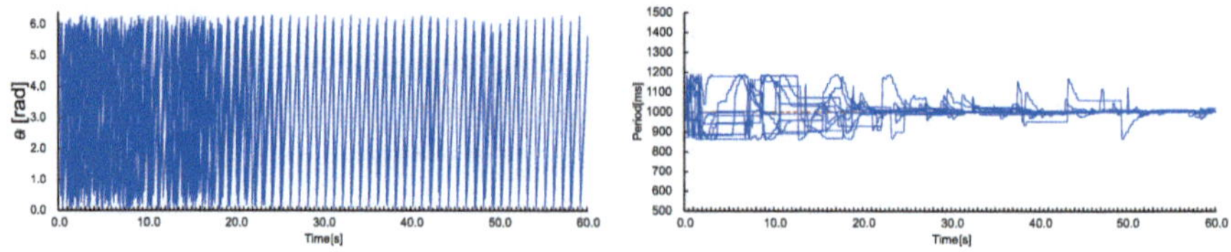

(a) Phase $\theta_i(t)$

(b) Measurement period. The red dashed line at 1 s represents the initial measurement period.

Fig. 5. Simulation results of the proposed synchronization algorithm with 50 mobile sensor nodes. The results of 10 out of 50 nodes are presented.

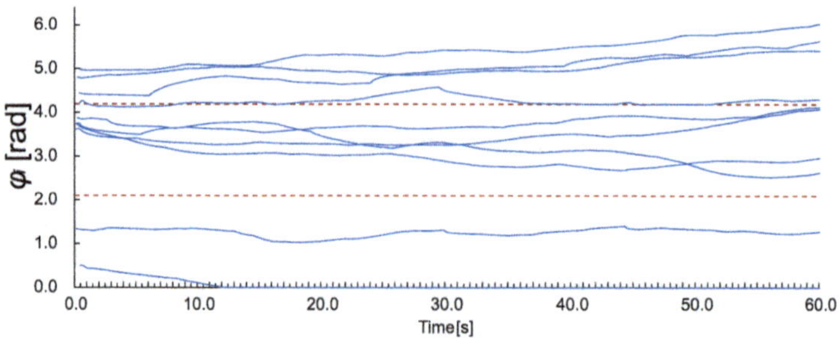

Fig. 6. Simulation results of the proposed desynchronization algorithm with 50 mobile sensor nodes when D is set to 3. The results of 10 out of 50 nodes are presented.

5 Related Works

5.1 Synchronization

Synchronization means performing periodic tasks of multiple sensor nodes simultaneously, as illustrated in Fig. 1. Synchronization can be observed in natural phenomena, for example, the collective flashing of fireflies [7]. Synchronization is not the same as time synchronization. Time synchronization aims to synchronize local clocks embedded in sensor nodes. However, when only the execution timings of periodic tasks are important, time synchronization is not always necessary, and synchronization, which ensures synchronization of the execution timings of periodic tasks, is sufficient.

Several mathematical synchronization models have been proposed [4,5]. Such mathematical synchronization models assume that a node receives messages instantly sent from another node. However, we cannot ignore communication delays in practical sensor networks. Therefore, we should take communication delays into account when proposing synchronization models for practical sensor networks. G. Werner-Allen et al. proposed a synchronization method considering realistic communication delays in sensor networks [9].

5.2 Desynchronization

Desynchronization means performing periodic tasks at the same frequency among multiple sensor nodes while offsetting their timings, as shown in Fig. 2. Desynchronization is useful for scheduling periodic tasks to manage resource allocation and avoid wireless communication congestion.

Several models have also been proposed [2,10]. Some works proposed models inspired by biological phenomena such as the calls of frogs [1,6]. Male Japanese tree frogs exhibit a behavior where they adjust their calling timings to avoid overlapping with others, ensuring their calls are more effectively heard by females. They modeled this behavior and proposed using the model to adjust wireless communication timings to avoid contention with other nodes.

Furthermore, most of the existing desynchronization algorithms aim to avoid wireless communication congestion. Few have focused on the coordination of periodic measurement timing. Existing desynchronization algorithms only require ensuring that communication timings do not overlap. In contrast, when coordinating measurement timings, the consistency of the measurement period may be critical. For these reasons, existing models are unsuitable for coordinating measurement timings in sensor networks.

6 Conclusion

We proposed two algorithms for the self-configured coordination of measurement timing in sensor networks, inspired by biological phenomena: a synchronization algorithm and a desynchronization algorithm. The synchronization algorithm ensures that sensor nodes measure their targets at the same time. In contrast, the desynchronization algorithm ensures that sensor nodes measure their targets at the same frequency while evenly offsetting their measurement phases among sensor nodes.

We evaluated the proposed algorithms with mobile sensor nodes through simulations. We demonstrated that the synchronization algorithm synchronized the periods and phases of measurement timings. Additionally, we showed that the desynchronization algorithm synchronized the periods of measurement timings, but it offset the phases.

For future work, we considered communication delays in our simulations, but we did not address the problems caused by packet loss, which occurs in practical wireless sensor networks. Therefore, it will be important to consider the impact of packet loss.

References

1. Aihara, I., Kominami, D., Hirano, Y., Murata, M.: Mathematical modelling and application of frog choruses as an autonomous distributed communication system. Roy. Soc. Open Sci. **6**(1) (2019). https://doi.org/10.1098/rsos.181117
2. Degesys, J., Rose, I., Patel, A., Nagpal, R.: DESYNC: self-organizing desynchronization and TDMA on wireless sensor networks. In: Proceedings of the 6th International Conference on Information Processing in Sensor Networks, IPSN07, pp. 11–20. Association for Computing Machinery, Massachusetts, Cambridge, USA (2007). https://doi.org/10.1145/1236360.1236363
3. Kandris, D., Nakas, C., Vomvas, D., Koulouras, G.: Applications of wireless sensor networks: an up-to-date survey. Appl. Syst. Innov. **3**(1), 14 (2020). https://doi.org/10.3390/asi3010014
4. Kuramoto, Y.: Chemical Oscillations, Waves and Turbulence. Springer, Berlin (1983)
5. Mirollo, R.E., Strogatz, S.H.: Synchronization of pulse-coupled biological oscillators. SIAM J. Appl. Math. **50**(6), 1645–1662 (1990)
6. Ron, D., Lee, J.R.: Fair and distributed resource allocation in wireless networks using frogs' calling behavior algorithm. IEEE Access **7**, 87313–87322 (2019). https://doi.org/10.1109/ACCESS.2019.2919130
7. Smith, H.M.: Synchronous flashing of fireflies. Science **82**(2120), 151–152 (1935). https://doi.org/10.1126/science.82.2120.151
8. Wang, Y., Doyle, F.J.: Exponential synchronization rate of Kuramoto oscillators in the presence of a pacemaker. IEEE Trans. Autom. Control **58**(4), 989–994 (2013). https://doi.org/10.1109/TAC.2012.2215772
9. Werner-Allen, G., Tewari, G., Patel, A., Welsh, M., Nagpal, R.: Firefly-inspired sensor network synchronicity with realistic radio effects. In: Proceedings of the 3rd International Conference on Embedded Networked Sensor Systems, SenSys 2005, pp. 142–153. Association for Computing Machinery, San Diego, California, USA, November 2005. https://doi.org/10.1145/1098918.1098934
10. Yu, U.S., Choi, H.H., Lee, J.R.: Kuramoto-desync: distributed and fair resource allocation in a wireless network. IEEE Access **7**, 104769–104776 (2019). https://doi.org/10.1109/ACCESS.2019.2932425

Real-Time Context-Aware Privacy Adaptation for IoT: A Lightweight Approach

Yaqin Yaseen Shaheen[1,2](✉), Miguel J. Hornos[2], and Carlos Rodríguez-Domínguez[2]

[1] Palestine Technical University-Kadoorie (PTUK), Tulkarm, Palestine
yaqin.shaheen@ptuk.edu.ps, yshaheen@correo.ugr.es
[2] Software Engineering Department, Research Centre for Information and Communication Technologies (CITIC-UGR), University of Granada, Aynadamar Campus, 18071 Granada, Spain

{mhornos,carlosrodriguez}@ugr.es

Abstract. With the increasing adoption of Internet of Things (IoT) devices in smart homes, wearables, and public environments, privacy concerns have emerged as a significant challenge. Traditional privacy mechanisms often rely on static policies or machine learning models that require high computational power, making them unsuitable for resource-constrained IoT devices. This paper presents a novel real-time, context-aware privacy adaptation approach that dynamically adjusts privacy settings based on multiple contextual factors, including user presence, location, and network security status. Unlike existing solutions that either focus solely on access control or rely on computationally intensive models, our rule-based approach provides a lightweight, fast, and developer-friendly solution. Our results demonstrate improved privacy protection while ensuring computational efficiency, making it well-suited for IoT applications in smart homes, wearables, and other resource-constrained environments.

Keywords: Internet of Things (IoT) · Privacy · Security · Context-Awareness · Context-Aware Security (CAS)

1 Introduction

The rapid proliferation of the Internet of Things (IoT) has significantly impacted diverse domains, such as smart homes, healthcare, industrial automation, and intelligent transportation. By interconnecting billions of devices, IoT facilitates seamless communication and automation, significantly transforming everyday activities. However, the heterogeneity and limited resources of IoT devices pose substantial security and privacy challenges. Traditional static security measures, which fail to adapt to the dynamic needs of IoT environments, leave systems vulnerable to a range of cyber threats, including unauthorized access, data breaches, and denial-of-service attacks [1].

Existing security solutions, while addressing some of these challenges, often struggle to meet the adaptive and resource-efficient requirements of IoT systems. Context-Aware

M. Rodrigues et al. (Eds.): ISAmI 2025, LNNS 1776, pp. 41–51, 2026.
https://doi.org/10.1007/978-3-032-14138-5_5

Security (CAS) has emerged as a promising solution, offering dynamic adaptation of security and privacy mechanisms based on contextual factors such as user behavior, device status, and environmental conditions. The term "context" in CAS refers to several elements that impact the system's security and privacy decisions, including a user's location, behavior, network conditions, and device state [2]. IoT systems can react instantly to changing threats thanks to this flexibility. However, many of the CAS solutions available today rely on computationally demanding models, such as artificial intelligence (AI) and machine learning (ML), which may have a substantial impact on data privacy and performance efficiency. In response, decentralized technologies like Federated Learning (FL) and Blockchain have drawn interest as supplementary ways to improve CAS frameworks by boosting data security and trust [3, 4]. However, deploying these technologies in resource-constrained IoT environments remains a challenge.

Consequently, there is an urgent need for computationally efficient and lightweight methods that can preserve security and privacy without overwhelming the limited resources of IoT devices.

In order to fill this gap, this study suggests a real-time, lightweight privacy adaptation method that dynamically modifies privacy settings according to contextual variables including user presence, location, and network security status. Unlike conventional access control techniques or AI-based solutions, our rule-based strategy ensures faster processing and seamless integration into existing IoT ecosystems. Our solution's ability to dynamically modify privacy in response to contextual information—like shifts in location or network security—without requiring constant user input is one of its primary features.

The remainder of this paper is organized as follows: Sect. 2 presents background information. Section 3 discusses key privacy challenges in IoT. Section 4 reviews related work, while Sect. 5 provides recommendations based on this analysis. Section 6 validates our proposed approach. Finally, Sect. 7 concludes the paper and suggests future research directions.

2 Background

IoT has revolutionized sectors such as healthcare, smart homes, industrial automation, and public infrastructure by embedding digital intelligence into everyday environments. The exponential growth of IoT has resulted in a proliferation of connected devices, generating massive amounts of data and necessitating robust security and privacy measures [5]. However, traditional security mechanisms, which rely on static policies, struggle to address the dynamic nature of IoT environments, leaving systems vulnerable to cyber threats, including unauthorized access, data breaches, and denial-of-service attacks [1, 6].

CAS has emerged as a promising solution by dynamically adapting security protocols based on real-time contextual factors, such as device status, user behavior, and environmental conditions. By enabling systems to respond dynamically to evolving threats, CAS offers a more flexible and adaptive approach than traditional static security models. However, many CAS implementations rely on computationally intensive technologies like AI and ML, which pose challenges in resource-constrained IoT environments [2, 7].

Decentralized technologies such as Blockchain and FL are explored to enhance the resilience and trustworthiness of CAS frameworks by providing decentralized data security. Despite their potential, implementing these technologies in IoT systems remains challenging due to the limited computational capabilities of many IoT devices [8].

Privacy protection is a critical concern in IoT, where continuous collection and transmission of sensitive user data increase the risk of breaches. Existing privacy-preserving techniques, including homomorphic encryption and FL, are often limited by the constrained resources of IoT devices. Context-aware privacy adaptation models offer a more dynamic and resource-efficient solution by adjusting privacy settings in response to contextual factors like user location, device status, and network conditions [9, 10].

Despite significant advancements in both security and privacy-preserving technologies, IoT environments continue to face challenges that hinder secure deployment. The next section will explore these challenges in detail, focusing on the limitations of current approaches and the need for more resource-efficient solutions.

3 Major IoT Privacy Challenges

The widespread adoption of IoT devices has significantly increased concerns about user privacy. IoT systems continuously collect and process vast amounts of personal, behavioral, and contextual information. This data exposure introduces several privacy challenges that must be addressed to ensure secure and ethical IoT deployment. The primary privacy concerns in IoT environments stem from resource constraints, data sensitivity, and the dynamic nature of IoT systems. As a result, all relevant entities in IoT ecosystems must prioritize this issue. The key privacy concerns include:

Data Sensitivity and Unauthorized Access. IoT devices continuously collect and transmit sensitive user information, such as health metrics, location data, and activity patterns. If inadequately protected, this data becomes vulnerable to unauthorized access, potentially leading to privacy breaches and misuse. Conventional security measures, such as authentication and encryption, often fail in dynamic IoT environments where data is continuously transferred across multiple platforms and devices. Moreover, combining data from various sources increases the risk of privacy violations, as attackers may infer sensitive information beyond what is expressly disclosed [3, 11].

Context Awareness and Privacy Risks. CAS systems adjust security measures in response to real-time factors, such as user location, social interactions, and environmental conditions, to provide adaptive protection. However, continuous data collection and processing may pose serious privacy risks. Violations can occur if users are unaware of how their contextual data is used. While context-aware methods enhance security, they must do so without disclosing private information or violating user privacy [2, 12].

Lack of User Control and Transparency. Limited user control over personal data is a major privacy issue in IoT. Many IoT devices have default settings that favor data collection, offering users few options to manage privacy preferences. The risk of misuse is further heightened by a lack of transparency in how data is processed, stored, and shared. To address this, user-centric privacy measures must enable individuals to dynamically define their data-sharing preferences [13].

Regulatory Compliance and Legal Challenges. IoT systems must comply with increasing data privacy laws, such as the California Consumer Privacy Act (CCPA) and the General Data Protection Regulation (GDPR) [9]. Ensuring compliance in resource-constrained IoT environments remains challenging. Automated privacy enforcement techniques must be integrated to maintain efficiency while adhering to regulations. However, the complexity and regional variations of privacy laws hinder the development of standardized security frameworks [14, 15].

Privacy-Preserving Mechanisms and Performance Constraints. Privacy-preserving techniques like FL, differential privacy, and homomorphic encryption aim to mitigate IoT privacy risks. However, these methods often introduce processing overhead, which is unfeasible for resource-limited IoT devices. Balancing security, privacy, and computational efficiency remains a challenge in IoT systems [16].

Despite advances in privacy-preserving technologies, IoT environments still face significant challenges due to their dynamic, heterogeneous, and resource-constrained nature. Many existing solutions rely on computationally demanding models or static regulations, which are impractical for low-power IoT devices. This underscores the urgent need for real-time, lightweight privacy adaptation systems capable of dynamically adjusting privacy settings without imposing excessive computational burdens.

4 Analysis of the Most Relevant Related Literature

The implementation of CAS in IoT has been the subject of numerous studies, with an emphasis on privacy protection, access control, and authentication. Perera et al. [7] examined the security and privacy challenges in context-aware systems, emphasizing the importance of securing contextual data across various layers, from collection to access control. Similarly, Li et al. [17] suggested a middleware-based security approach that inhibits unwanted access to contextual data by using authentication and encryption methods. Their results show that cryptographic techniques enhance security but also highlight computational challenges, especially in resource-limited IoT devices.

Additional research on CAS applications has been conducted in particular IoT fields. Sylla et al. [2] examined CAS methods in smart cities, demonstrating how context-aware authentication and access control enhance security. Meanwhile, Chatterjee et al. [18] examined machine-to-machine authentication in industrial IoT and suggested frameworks that reduce human involvement while maintaining strong security. However, their research focuses mainly on industrial settings, leaving a gap in CAS applications for other IoT domains.

Several studies have integrated ML and AI into CAS models. Kayes et al. [19] explored AI-based security frameworks, recognizing their potential to improve real-time security decision-making. However, data privacy, energy efficiency, and computational complexity remain major challenges. Grimm et al. [9] analyzed CAS in vehicular networks, presenting a taxonomy of security requirements and emphasizing the challenges associated with real-time context processing.

Despite the breadth of research on CAS in IoT security, existing studies often focus on isolated aspects, such as authentication or access control, without offering a holistic view of CAS models. Additionally, scalability and adaptability remain key challenges, particularly in large-scale IoT deployments. While CAS solutions that integrate ML and cryptographic privacy techniques offer strong security guarantees, they introduce high computational costs, which can be impractical for resource-limited IoT devices.

This paper proposes a lightweight, real-time privacy adaptation framework that dynamically adjusts security and privacy settings. Unlike existing models that rely heavily on computationally expensive AI-based techniques, our approach uses a rule-based framework to provide enhanced privacy protection while ensuring low-latency performance in IoT ecosystems. The discussion below delves into these aspects:

Adaptive and Dynamic Privacy in CAS. CAS privacy mechanisms provide flexible security solutions that adjust in real-time to contextual factors. Dynamic security policy enforcement adapts measures based on factors such as device location, network conditions, user behavior, and risk levels [20]. This flexible approach ensures data encryption and access control procedures remain effective despite environmental changes.

Protecting privacy through adaptive risk inference—which entails dynamically evaluating privacy concerns and modifying security settings accordingly—is another crucial component. Research emphasizes the function of CAS models in striking a balance between data security and usability by automatically altering data-sharing guidelines according to user context and sensitivity levels [21].

Regulatory Compliance and Privacy Adaptation. The CCPA and GDPR play a crucial role in shaping IoT privacy policies as concerns grow. To guarantee that IoT systems are both safe and adhere to international privacy requirements, these legal frameworks must be incorporated into CAS models. Our approach ensures data handling aligns with user consent and regulations, addressing this gap.

5 Proposed Approach and Key Analysis Recommendations

This study addresses growing IoT privacy concerns, particularly in resource-limited settings like wearables and smart homes. Our real-time, context-aware privacy adaptation strategy provides a dynamic alternative to static privacy policies and computationally intensive ML models. It improves privacy while preserving low computational overhead by adjusting privacy settings based on various contextual factors, such as user presence, location, and network security. This approach is designed to address the limitations of existing privacy solutions, which struggle to scale in low-power IoT devices.

5.1 Features of Our Approach

The proposed framework incorporates key features that enable efficient, adaptable, and privacy-aware operation in resource-constrained IoT environments.

Real-Time Context-Aware Privacy Adaptation. Our method dynamically adjusts privacy settings based on network security, user activity, and location. This real-time adaptation enables continuous monitoring and policy updates, ensuring effective privacy protection as circumstances change. This approach improves privacy enforcement by proactively responding to changing security contexts, in contrast to static policies.

Lightweight Rule-Based System. The suggested rule-based solution avoids the high processing needs of AI and ML models while guaranteeing computational efficiency. It is perfect for resource-constrained IoT devices like wearables and smart homes because of its lightweight architecture.

5.2 Privacy Adaptation Process

Figure 1 presents a structured flowchart illustrating how our approach dynamically adjusts privacy settings based on contextual factors. The process includes the following stages:

1. **Context Detection.** The system continuously collects user context data, including user presence, location, and network security status, through IoT sensors and network monitoring.
2. **Privacy Rule Matching.** The collected contextual data is evaluated against predefined privacy rules to determine the necessary privacy action.
3. **Privacy Enforcement.** Based on rule evaluation, the system automatically applies the appropriate privacy action. This may include *muting the microphone* (when a guest is detected), *disabling the camera* (when connected to an open Wi-Fi network), and *restricting data sharing* (when the user is in public locations).

This rule-based adaptation ensures lightweight, efficient privacy enforcement, making it suitable for resource-constrained IoT environments.

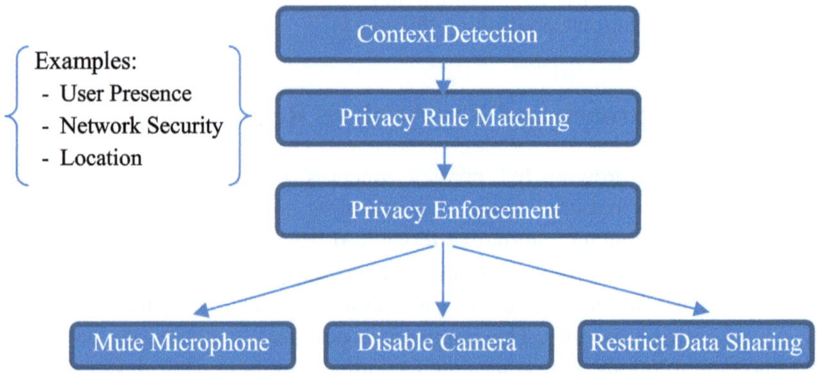

Fig. 1. Flowchart of the proposed real-time context-aware privacy adaptation method.

5.3 Proposed Recommendations from Analysis

Based on our findings, we outline the following key recommendations to improve the effectiveness and scalability of the proposed approach.

Integration of Regulatory Compliance. While our approach enhances real-time privacy adaptation, integrating compliance features like GDPR and CCPA could further improve real-world applicability. This includes user consent management, data retention, and the "right to be forgotten," making it more versatile and industry-compliant.

Expansion of Contextual Factors for Privacy Adaptation. To enhance the framework's comprehensiveness, incorporating additional contextual factors—such as device security, user risk profiles, and network bandwidth—can improve the precision of privacy decisions. Considering both real-time context and long-term usage patterns, the system ensures more personalized and robust privacy protection. This expansion supports diverse IoT applications, addressing multi-layered privacy needs across various use cases.

Scalability for Large-Scale IoT Deployments. While effective in small-scale applications, scaling the approach to larger IoT environments (e.g., industrial settings, smart cities) is a crucial next step. Incorporating techniques such as cloud/edge computing and distributed processing would enable the system to handle a high volume of devices and users. Cloud computing, while powerful, suffers from high latency, bandwidth limitations, and centralized privacy risks. Edge computing improves latency by enabling local data processing, but introduces new challenges such as a broader attack surface and more complex endpoint security [22].

User-Centric Customization of Privacy Policies. Future improvements could involve enabling user-centric customization of privacy settings. While the system auto-adjusts privacy settings, enabling user overrides could boost trust and enhance data control.

6 Validation of the Proposed Approach

To evaluate the effectiveness of our real-time, context-aware privacy adaptation framework, we implemented a Python-based simulation using Flask and Matplotlib. The simulation tested real-time privacy adaptation based on a range of contextual factors, including user presence, location, and network security, offering a comprehensive view of its adaptability. These factors were explicitly categorized to illustrate the system's behavior across different scenarios, ensuring a clear understanding of its performance in diverse conditions. The results demonstrated that the system achieved an average response time of less than 2 s, ensuring efficient privacy enforcement. This suggests the framework is computationally efficient and suitable for IoT in resource-constrained environments under the tested conditions. However, further validation across diverse real-world scenarios is needed to assess its potential and robustness.

The simulation results further support the need for a dynamic, context-aware privacy adaptation approach in IoT systems. As shown in Fig. 2, the distribution of privacy levels highlights that the framework often enforces high privacy settings in response to elevated contextual risks—such as when the device is in a public setting or connected to an insecure network. This minimizes data exposure while dynamically relaxing privacy in secure environments for a seamless yet secure user experience. The frequent enforcement of the "High" privacy level reflects the system's conservative approach in potentially risky contexts. In contrast to static policies, which may either limit usability or fall short in protection, the framework adjusts in real time. Its ability to apply lower privacy levels in safer conditions highlights both its flexibility and practical efficiency. This distribution also suggests that the system consistently prioritizes risk mitigation, aligning privacy enforcement with context severity rather than applying uniform rules. This adaptability is especially beneficial in low-power IoT environments, where balancing protection and functionality is essential.

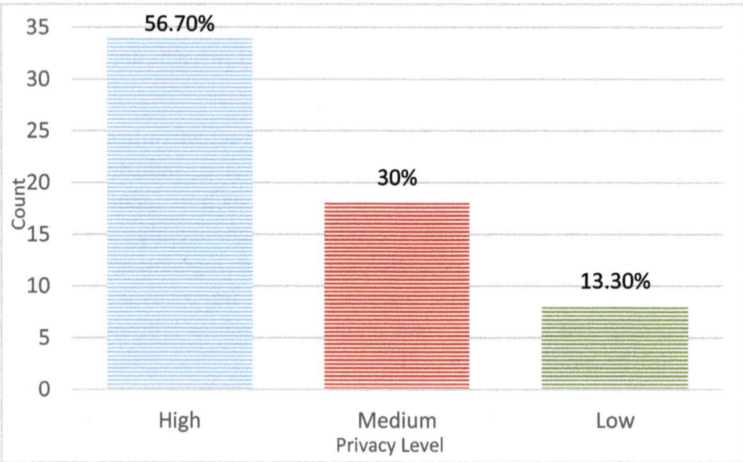

Fig. 2. Distribution of privacy levels.

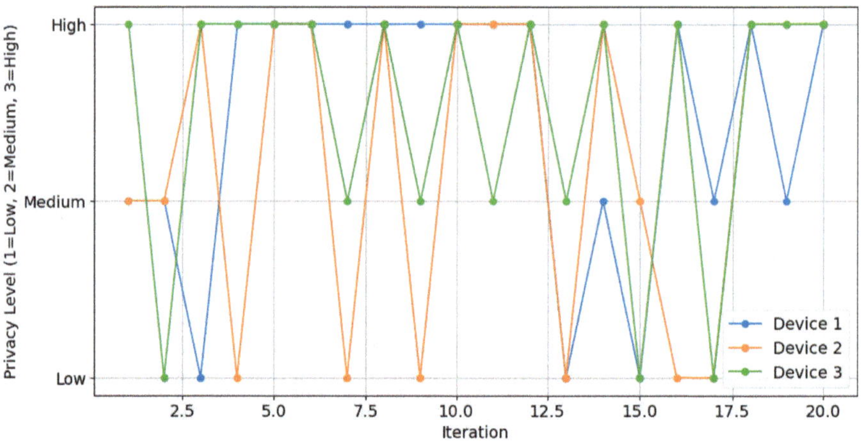

Fig. 3. Privacy level adaptation over time.

Additionally, the privacy adaptation trends over time, illustrated in Fig. 3, demonstrate the system's capacity to dynamically respond to real-time contextual changes. Different devices show distinct fluctuations in privacy levels, reflecting how environmental factors influence privacy enforcement. This confirms that the framework effectively balances privacy and usability—an advantage over static policies. Variation in adaptation across devices highlights the framework's context-aware and device-specific behavior, avoiding one-size-fits-all thresholds. This targeted enforcement reduces unnecessary restrictions while preserving user safety—crucial in dynamic IoT contexts that demand prompt and effective responses.

7 Conclusions and Future Work

The rapid growth of IoT devices has highlighted the urgent need for efficient privacy-preserving systems, particularly in resource-constrained environments. Current CAS solutions often fail to balance computational efficiency and adaptability, limiting their applicability in large-scale IoT deployments.

This paper proposed a new lightweight approach for real-time privacy adaptation that dynamically adjusts privacy settings based on key contextual factors such as user presence, location, and network security status. Unlike computationally intensive approaches, our rule-based methodology ensures rapid decision-making while minimizing computational overhead.

This work contributes to the growing body of research by addressing the limitations of existing privacy-preserving techniques in IoT, and highlighting the need for adaptive, resource-efficient solutions that scale across diverse applications. Future research will focus on expanding the contextual parameters, improving scalability, and exploring automation of regulatory compliance to facilitate real-world IoT integration. We also plan to evaluate the framework in real-world settings—such as smart homes and healthcare systems—to assess its practical scalability. Furthermore, we are considering a distributed rule engine to enhance performance in edge-based deployments. To improve regulatory compliance, we plan to integrate automated mapping of contextual data to privacy regulations such as GDPR. These enhancements will support broader adoption of the approach in complex and dynamic IoT environments.

Acknowledgments. The authors would like to thank Palestine Technical University - Kadoorie for their financial support in conducting this research. This work was also supported by grants PID2023-149185OB-I00, PID2022-139297OB-I00, and TED2021-132262A-I00, all funded by MICIU/AEI/10.13039/501100011033, that is, the Spanish Ministry of Science, Innovation and Universities (State Research Agency), and co-financed by ERDF/EU (for the first two grants) and by NextGenerationEU (for the latter).

Disclosure of Interests. The authors declare no competing interests relevant to the content of this article.

References

1. Butun, I., Österberg, P., Song, H.: Security of the Internet of Things: vulnerabilities, attacks, and countermeasures. IEEE Commun. Surv. Tutor. **22**, 616–644 (2020). https://doi.org/10.1109/COMST.2019.2953364
2. Sylla, T., Chalouf, M.A., Francine, K., Samake, K.: Context-aware security in the Internet of Things: a survey. Int. J. Auton. Adapt. Commun. Syst. **14**, 1 (2021). https://doi.org/10.1504/IJAACS.2021.10035751
3. Anand, P., Singh, Y., Selwal, A., Alazab, M., Tanwar, S., Kumar, N.: IoT vulnerability assessment for sustainable computing: threats, current solutions, and open challenges. IEEE Access **8**, 168825–168853 (2020). https://doi.org/10.1109/ACCESS.2020.3022842

4. Gugueoth, V., Safavat, S., Shetty, S.: Security of Internet of Things (IoT) using federated learning and deep learning — recent advancements, issues and prospects. ICT Express **9**, 941–960 (2023). https://doi.org/10.1016/j.icte.2023.03.006
5. Shaheen, Y., Hornos, M.J., Rodríguez-Domínguez, C.: IoT security and privacy challenges from the developer perspective. In: Novais, P., et al. (eds.) ISAmI 2023. LNNS, vol. 770. Springer, Cham (2023). https://doi.org/10.1007/978-3-031-43461-7_2
6. Makhdoom, I., Abolhasan, M., Lipman, J., Liu, R.P., Ni, W.: Anatomy of threats to the Internet of Things. IEEE Commun. Surv. Tutor. **21**, 1636–1675 (2019). https://doi.org/10.1109/COMST.2018.2874978
7. Perera, C., Zaslavsky, A., Christen, P., Georgakopoulos, D.: Context aware computing for the Internet of Things: a survey. IEEE Commun. Surv. Tutor. **16**, 414–454 (2014). https://doi.org/10.1109/SURV.2013.042313.00197
8. Ge, C., Liu, Z., Fang, L.: A blockchain based decentralized data security mechanism for the Internet of Things. J. Parallel Distrib. Comput. **141**, 1–9 (2020). https://doi.org/10.1016/j.jpdc.2020.03.005
9. Grimm, D., Stang, M., Sax, E.: Context-aware security for vehicles and fleets: a survey. IEEE Access **9**, 101809–101846 (2021). https://doi.org/10.1109/ACCESS.2021.3097146
10. Torre, D., Chennamaneni, A., Rodriguez, A.: Privacy-preservation techniques for IoT devices: a systematic mapping study. IEEE Access **11**, 16323–16345 (2023). https://doi.org/10.1109/ACCESS.2023.3245524
11. Fazeldehkordi, E., Grønli, T.-M.: A survey of security architectures for edge computing-based IoT (2022). https://doi.org/10.3390/iot3030019
12. Corallo, A., Lazoi, M., Lezzi, M., Luperto, A.: Cybersecurity awareness in the context of the Industrial Internet of Things: a systematic literature review. Comput. Ind. **137**, 103614 (2022). https://doi.org/10.1016/j.compind.2022.103614
13. Pinto, G.P., Donta, P.K., Dustdar, S., Prazeres, C.: A systematic review on privacy-aware IoT personal data stores (2024). https://doi.org/10.3390/s24072197
14. Aljeraisy, A., Barati, M., Rana, O., Perera, C.: Privacy laws and privacy by design schemes for the Internet of Things: a developer's perspective. ACM Comput. Surv. **54**, 1–38 (2021). https://doi.org/10.1145/3450965
15. Shaheen, Y.Y., Hornos, M.J., Rodríguez-Domínguez, C.: Addressing privacy challenges in Internet of Things (IoT) applications. In: Novais, P., et al. (eds.) ISAmI 2024. LNNS, vol., 1279, pp. 45–54. Springer, Cham (2025). https://doi.org/10.1007/978-3-031-83117-1_5
16. Wakili, A., Bakkali, S.: Privacy-preserving security of IoT networks: a comparative analysis of methods and applications. Cyber Secur. Appl. **3**, 100084 (2025). https://doi.org/10.1016/j.csa.2025.100084
17. Li, X., Eckert, M., Martinez, J.-F., Rubio, G.: Context aware middleware architectures: survey and challenges (2015). https://doi.org/10.3390/s150820570
18. Chatterjee, B., Cao, N., Raychowdhury, A., Sen, S.: Context-aware intelligence in resource-constrained IoT nodes: opportunities and challenges. IEEE Des. Test **36**, 7–40 (2019). https://doi.org/10.1109/MDAT.2019.2899334
19. Kayes, A.S.M., et al.: A survey of context-aware access control mechanisms for cloud and fog networks: taxonomy and open research issues (2020). https://doi.org/10.3390/s20092464
20. Bou Chaaya, K., Barhamgi, M., Chbeir, R., Arnould, P., Benslimane, D.: Context-aware system for dynamic privacy risk inference: application to smart IoT environments. Future Gener. Comput. Syst. **101**, 1096–1111 (2019). https://doi.org/10.1016/j.future.2019.07.011

21. Czekster, R.M., Webber, T., Furstenau, L.B., Marcon, C.: Dynamic risk assessment approach for analysing cyber security events in medical IoT networks. Internet Things **29**, 101437 (2025). https://doi.org/10.1016/j.iot.2024.101437
22. Singh, P., Gautam, P., Arya, N.: Role of edge computing in IoT. In: 2025 IEEE 14th International Conference on Communication Systems and Network Technologies (CSNT), pp. 1226–1230 (2025). https://doi.org/10.1109/CSNT64827.2025.10968010

Smart Home Control Through Sign Language Recognition

Antonio De Toro-Castro[1], Marcos Lupión[1,2](✉), Vicente González-Ruiz[1], Juan Francisco Sanjuan[1], and Pilar M. Ortigosa[1]

[1] Department of Informatics, ceiA3, Univ. of Almería, 04120 Almería, Spain
adc144@inlumine.ual.es, {marcoslupion,vruiz,jsanjuan,ortigosa}@ual.es
[2] School of Computing, Ulster University, Belfast BT15 1ED, UK
M.LupionLorente@ulster.ac.uk

Abstract. Sign Language Recognition (SLR) plays a crucial role in improving accessibility for individuals with hearing impairments, particularly within Smart Home environments. This paper presents Smart-LSE-UAL, a novel real-time Spanish Sign Language (LSE) recognition system designed to facilitate interaction with smart home devices. The proposed system leverages a hybrid CNN-LSTM neural network architecture to process real-time videos. The keypoints required for recognition (hands and body) are extracted using MediaPipe, ensuring a compact and privacy-preserving representation of user gestures. A newly recorded LSE dataset, specifically tailored for smart home control, is introduced to enhance recognition accuracy. The system is fully integrated with a KNX-based smart home infrastructure at the University of Almería, enabling seamless control of household components, such as lights, shutters, and air conditioning, via sign language commands. Experimental evaluations demonstrate the effectiveness of the proposed approach, achieving high accuracy in gesture recognition for different users, while maintaining real-time performance.

Keywords: Language sign recognition · Smart Environments · Artificial Intelligence

1 Introduction

Hearing impairment, whether partial or total, imposes significant barriers to effective communication. In Europe, it is estimated that around 190 million people—approximately 20% of the population—experience some level of hearing loss or deafness [16]. Although many individuals with hearing loss benefit from assistive technologies like hearing aids and cochlear implants to enhance auditory input, a distinct subset of the deaf community, estimated at roughly 750,000 people in Europe, relies exclusively on sign language for communication [5]. Given the limited number of hearing individuals proficient in sign language

M. Rodrigues et al. (Eds.): ISAmI 2025, LNNS 1776, pp. 52–61, 2026.
https://doi.org/10.1007/978-3-032-14138-5_6

[5], researchers have developed automated Sign Language Recognition (SLR) systems that employ artificial intelligence (AI) and computer vision to interpret sign language from visual or sensor-based inputs [14]. These systems convert sign language into text or speech in real time, promising enhanced inclusivity and reduced reliance on human interpreters. SLR typically involves the segmentation of sign language into basic units, known as "glosses," each corresponding to a distinct word or phrase.

Due to their higher recognition accuracy and ease of deployment, vision-based approaches have gained increasing attention. Architectures such as convolutional neural networks (CNNs) [4] and Transformers [15] have proven effective in automatically extracting and fusing spatial and temporal features from video sequences, thus significantly enhancing recognition performance.

Ambient Intelligence within Smart Environments (SEs) is designed to assist individuals (particularly the elderly and those with disabilities or illnesses) in their daily activities [3]. SE applications include the recognition of activities of daily living (ADLs) [3], fall detection and assessment [11], and medication reminders [10]. Interaction with SEs is commonly facilitated by voice-controlled assistants, such as Amazon Alexa and Google Assistant, which have been shown to improve usability compared to traditional mobile or web interfaces [7]. For instance, fall detection systems may assess the condition of a user post-fall [12], while platforms like OpenHab or HomeAssistant enable voice-activated control of home devices.

Despite the clear need for inclusive technology, few SE solutions have been tailored to accommodate sign language users [8]. Efforts to merge SLR with SEs have focused on developing systems that perform real-time sign language interpretation using wearable or visual inputs [1]. One pioneering study in this field is presented in [9], where the authors designed a dynamic SLR system based on 3D CNNs for smart home interaction. Their system translates video sequences into home automation commands (e.g., "Turn ON Air conditioner in the bathroom") and is capable of recognizing up to 25 gestures from the Croatian Sign Language Dataset. While numerous studies report high recognition performance in controlled environments [14], many of these approaches are impractical for real-time deployment due to their heavy computational demands. Moreover, discrepancies between the perspectives in training datasets and those encountered in real-world applications can further compromise recognition accuracy.

In this paper, we introduce Smart-LSE-UAL, an innovative DL solution for real-time Spanish Sign Language recognition within the Smart Home environment at the University of Almería. The system leverages a CNN-LSTM model to process video sequences and accurately interpret sign language commands. In parallel, when recognising the gestures, a module dynamically controls key Smart Home components such as windows, shutters, and air conditioning, ensuring seamless interaction with the environment. The model is trained on a LSE dataset specifically created to include vocabulary and commands relevant to Smart Home operations.

The principal contributions of this work are as follows:

- The development of a robust, real-time Spanish Sign Language recognition system for smart environments.
- The creation of a novel LSE dataset specifically designed for Smart Home interaction commands.
- The implementation of a hybrid CNN-LSTM model optimized for deployment on resource-constrained devices.

The remainder of this paper is organized as follows. Section 2 details the proposed methodology, Sect. 3 describes the experimental setup and presents the evaluation results, and Sect. 4 concludes the paper with a discussion of future research directions.

2 Methodology

This section thoroughly describes the architecture and core components of the system.

2.1 Architecture

The system architecture (shown in Fig. 1) consists of several interconnected components for efficient sign language translation and smart home control. Following, the components and their task in the overall systems are described:

- **Camera**. It captures real-time video of the user's sign language gestures. It is configured to run 4 Frames Per Second (FPS).
- **Window Splitting Module**. It divides the video into 12-frame sequences for temporal gesture analysis (3-s windows).
- **Pose Recognition Component (Mediapipe)**. Processes each frame to detect and extract three different types of keypoints (hands and body), simplifying the data into a compact form for efficient processing.
- **Image generation**. It organizes the extracted keypoints into 100 by 100 images with black backgrounds, structuring the AI model's sequence data.
- **AI Model**. It consists of the AI model that analyzes the batch of images, classifying the sequences into gestures.
- **Smart Home Integration**. A script summarizes the gestures recognized by the AI model and builds the commands that allow to control the different Smart Home components. The Spacelynk server receives the commands and triggers actions to control devices such as lights, windows, and air conditioners.

2.2 Keypoints Extraction

Keypoints detection is performed using MediaPipe models [2] to capture the spatial positions of essential landmarks in human pose estimation and hand tracking. The extracted keypoints serve as a fundamental input representation for downstream tasks such as gesture classification. We extract keypoints from two primary regions:

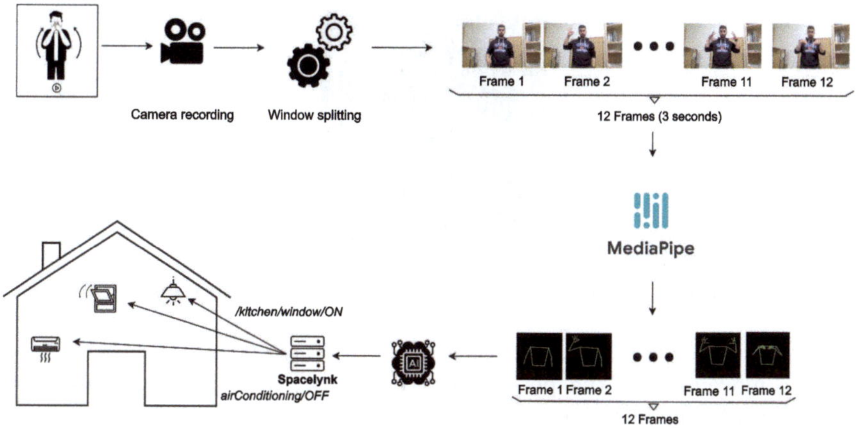

Fig. 1. System Architecture

- Pose: A subset of 8 upper-body landmarks for tracking arms and torso movements. These keypoints are selected to track the shoulders, elbows, wrists, and hips, which are essential for interpreting sign language gestures. The focus on upper body movements allows for efficient and accurate recognition of gestures involving arm positioning and torso orientation.
- Hands: Twenty-one keypoints covering the entire hand structure for precise gesture tracking. With 21 keypoints per hand, the total reaches 42 when both hands are detected. Hand gestures are central to sign language, and these keypoints track the wrist, finger positions, and joint movements, allowing for detailed and accurate gesture recognition. This level of detail enables the system to detect complex hand movements and finger configurations that are critical for understanding sign language.

Once these keypoints are extracted, they are used to create a 100×100 pixel image that accurately represents the sign speaker or user. This image is constructed by mapping the (x, y) coordinates of the detected keypoints onto a grid. The background is set to black to homogenize different backgrounds, and the keypoints are normalized, placing the user in the centre of the image to avoid changes in the scale of different users.

The result is a compact, privacy-preserving representation of the user, as it avoids using any personally identifiable features such as faces or full body images. Instead, the image keeps enough detail to recognize and classify sign language gestures while ensuring the privacy of the individual, as shown in Fig. 2.

2.3 Models

Batches of twelve images (captured in a three-second window) are processed in order to recognize a gesture on them. The input captures both spatial information (e.g., the person, the background, and location) and temporal relations

Fig. 2. Keypoint-based structural pose representation.

(e.g., the movement of the person across images). In this case, two different neural network configurations have been designed and evaluated:

- *CNN*: The CNN is particularly adept at capturing spatial relationships in images, thanks to its convolutional layers that can automatically extract spatial features. In this case, given that the input is a batch of data, each CNN layer is applied to each layer separately. A total of 5 convolutional layers are incorporated, being followed by a Global Average Pooling layer that flattens the data, and two fully connected layers that map the spatial features into the different gestures. Furthermore, dropout and batch normalization layers are incorporated to reduce overfitting and accelerate the training, respectively.
- *CNN + LSTM*: Since the input images also contain temporal information, we process image sequences (12 images) to recognize each gesture. Recurrent Neural Networks, particularly those incorporating LSTM units [6], are well-suited for temporal data processing. However, these networks cannot directly handle raw images, so we propose combining them with convolutional layers that first extract spatial features. For this reason, the CNN+LSTM architecture proposed is the same as the CNN described before but incorporates an LSTM layer with 12 units before the fully connected layers.

2.4 Smart Home Integration

The Smart Home at the University of Almería [13] is a laboratory specifically designed to facilitate research, experimentation, and educational activities related to ambient intelligence solutions. This environment enables both university staff and students to explore, develop, and learn about innovative smart home technologies. The facility is fully interconnected through a KNX bus system, which allows wired communication among various devices. A comprehensive map of the installed devices within the laboratory is presented in Fig. 3.

The system incorporates an array of sensors capable of monitoring environmental parameters such as humidity, temperature, and CO_2 levels. Additionally, actuators are integrated into the environment, including window blinds, air conditioning units, lighting systems, and shutters, allowing comprehensive automation and control functionalities.

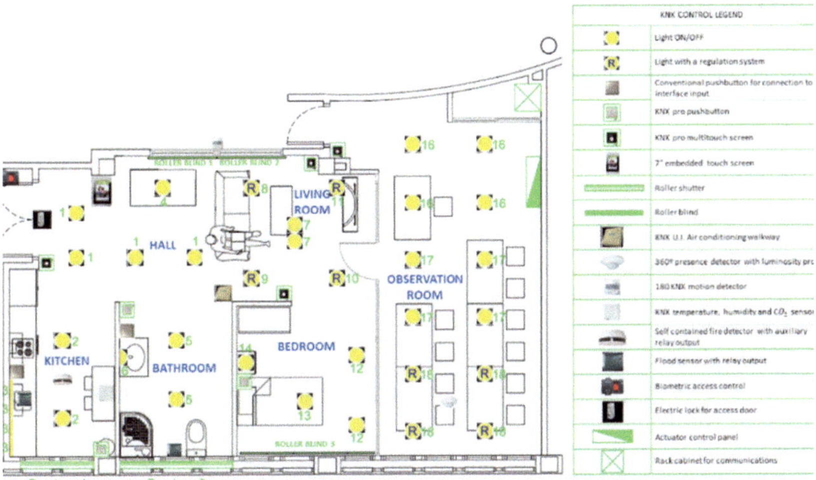

Fig. 3. Diagram of sensors and actuators in the Smart Home environment.

The central control system of the Smart Home is managed by a web server operating on the SpaceLynk LSS100200 platform. This server functions as the primary interface for the KNX system, allowing users to manage and monitor device statuses through either a web-based interface or a RESTful API. This API facilitates programmatic interactions via HTTP requests, enabling integration with external applications and automation scripts.

Each sensor and actuator within the system is uniquely identified by a device address structured in a three-level hierarchy (e.g., 3/1/1 corresponds to the kitchen light). Through the API, it is possible to both query the status of devices and modify their operational state.

The primary objective of this study is to implement device control using a Speech Language Recognition (SLR) system integrated with a Language-Specific Engine (LSE). The control logic relies on accurately identifying commands composed of four distinct elements. An API call is triggered only when all components of the command are correctly recognized. The command structure is defined as follows:

1. **Room**: *Kitchen, Bedroom, Bathroom, Living Room.*
2. **Item**: *Light, Shutter, Air Conditioning.*
3. **Action**: *Turn on, Turn off, Set to the middle (50%).*

This framework ensures precise and efficient interaction with the smart home system, facilitating natural language-based control while ensuring operational accuracy and minimizing potential errors in device activation.

3 Experimentation

In this section, the details of the proposed dataset and the experiments are provided.

3.1 Dataset

As mentioned in the introduction, there is no large LSE dataset that incorporates all the possible gestures that allow to control the Smart Home. For this reason, a new specific dataset has been recorded. Its details are the following:

- *Number of words*: 11.
- *Total number of sequences*: 110 per user. Augmented to 2750 for each user. 11000 in total.
- *Number of actors*: 4.
- *Number of locations*: 3.

The recording and annotation of the dataset is a very costly and time-consuming process. To automate the recording, a web tool has been designed. The main functionality can be appreciated in Fig. 4. In such a screen, the camera is activated and the web system requires the user to perform a gesture in a 3-s time window. At the right part, the ground truth gesture performed by a specialist actor is provided. Once the videos are recorded, these are uploaded to a server, executing the Mediapipe framework to extract the keypoints.

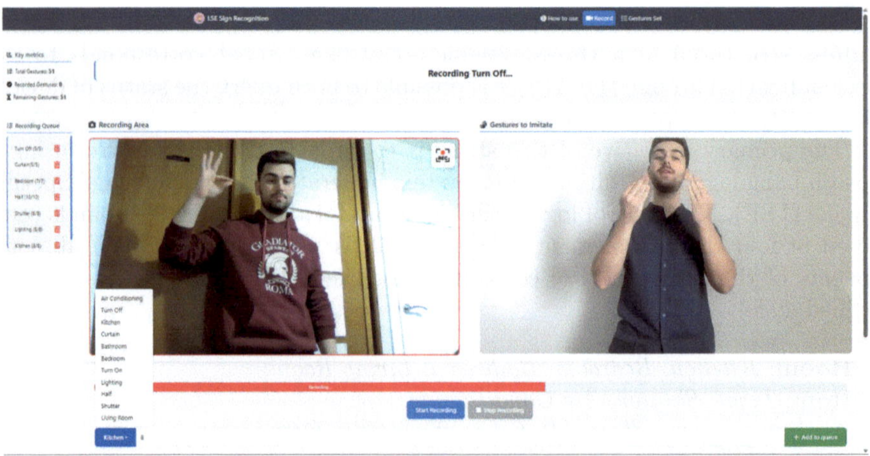

Fig. 4. Web tool to create the dataset.

3.2 Model Performance

In this section, the evaluation of the models is included. It has been carried out using a cross-validation Leave-One-Subject-Out approach. Taking into account the 4 users included in the dataset, the experimentation consisted of 4 training processes for each model. In each training, 3 users were used to train/validate the ANN, and one user for testing. This validation process can determine the generalization capabilities of the model with new and unseen users.

In the first experiment, two ANN architectures were compared: a CNN and a CNN-LSTM, which are described in Sect. 2.3. Results are shown in Table 1, which includes the training time, the number of epochs, and the train, validation and test accuracies of both models. It is important to notice that the different values are the average of the training for the different (4) users.

It can be appreciated that the CNN-LSTM model achieves a higher accuracy in validation and testing, which demonstrates that it is more convenient for adoption in the real environment. However, the training time (and the number of epochs) is slightly higher than the CNN model, which includes fewer parameters. It is also important to remark that training and validation datasets achieve 100% accuracy, while the test dataset (with the unseen user) achieves 87.93%, demonstrating the ability of the model to generalize with new users.

Table 1. Performance of model architectures.

Model	Training Time (s)	Num. epochs	Accuracy		
			Train	Validation	Testing
CNN	518	44	1.0000	0.9000	0.8145
CNN-LSTM	664	53	1.0000	1.0000	0.8793

Second, the f1-score metric of the gestures for the test dataset is included in Table 2. It is important to note that the 4 different trainings are included to appreciate the differences between users. It can be appreciated that mean f1-score values range from 0.71 (in the first user) to 0.87 in the case of the second user, with a mean value of 0.78. Although the data from the different users is balanced, the differences in the execution of the different gestures can produce such disparities in the accuracy of the model.

Regarding the different gestures, the maximum f1-score value is achieved by the "Air Conditioning" gesture, with an f1-score value of 0.91. It is important to remark that this gesture is very different from the others, being the only one that has hand movements at head height throughout the entire execution of the gesture. In contrast, the "Half" gesture achieves a lower f1-score of only 0.67 and fails to be recognized in the case of the first user. This poor performance is primarily due to its high visual similarity with the "Kitchen" gesture; both gestures involve similar hand positions and motion trajectories. Consequently, the model often confuses one with the other, leading to a misclassification of "Half" as "Kitchen".

Table 2. F1-score for the different users in test dataset.

Gesture		F1-Score				
ID	Meaning	1	2	3	4	Mean
1	Air Conditioning	0.9482	0.9879	0.9470	0.7924	0.9188
2	Turn Off	0.7368	0.7939	0.1011	0.8908	0.6306
3	Kitchen	0.6568	0.9718	0.7974	0.9859	0.8454
4	Curtain	0.8460	0.9443	0.7814	0.8208	0.8481
5	Bedroom	0.8556	0.8609	0.8861	0.8104	0.8532
6	Turn On	0.7205	0.7817	0.7778	0.8714	0.7878
7	Lighting	0.7893	0.9384	0.5569	0.7911	0.7689
8	Half	0.0000	0.9280	0.7911	0.9676	0.6716
9	Shutter	0.8039	0.8914	0.9153	0.8235	0.8585
10	Living Room	0.6737	0.7717	0.5476	0.9843	0.7443
11	Bathroom	0.7924	0.8000	0.8372	0.5994	0.7572
All		0.7112	0.8790	0.7217	0.8489	0.7894

4 Conclusions and Future Works

This work has introduced Smart-LSE-UAL, a real-time system for Spanish Sign Language recognition integrated within a KNX-based smart home environment. The proposed neural network combines spatial and temporal modelling via a CNN-LSTM architecture, using MediaPipe to extract pose and hand-based keypoints and transform them into compact, privacy-preserving image representations. The results obtained through cross-user validation demonstrate the model's ability to generalize across different individuals, achieving an average F1-score of 0.789 in new unseen users.

Despite these promising outcomes, certain gestures such as "Half" and "Turn Off" exhibit reduced recognition performance, due to high similarity between signs and variability in user execution. These findings highlight the need for improved training data and model refinement.

Future work will focus on several directions. Firstly, the dataset will be expanded to include a larger number of users, recording conditions, and sign vocabulary. This will not only increase the system's generalizability but also allow for more complex sentence-level commands. Secondly, efforts will be made to optimize the model for embedded deployment, ensuring energy-efficient inference in edge devices. Finally, usability studies with sign language users will be conducted to assess the practical impact of the system and identify additional accessibility needs.

Acknowledgements. This work has been funded by the projects R+D+i PID2021-123278OB-I00 and PDC2022-133370-I00 from MCI-N/AEI/10.13039/501100011033/ and ERDF funds; and the Department of Informatics of the University of Almería.

References

1. Adeyanju, I.A., Bello, O.O., Adegboye, M.A.: Machine learning methods for sign language recognition: a critical review and analysis. Intell. Syst. Appl. **12**, 200056 (2021)
2. Bazarevsky, V.: BlazePose: on-device real-time body pose tracking. arXiv preprint arXiv:2006.10204 (2020)
3. Chen, L., Nugent, C.D., Biswas, J., Hoey, J.: Activity Recognition in Pervasive Intelligent Environments, vol. 4. Springer Science & Business Media (2011)
4. Cui, R., Liu, H., Zhang, C.: A deep neural framework for continuous sign language recognition by iterative training. IEEE Trans. Multimedia **21**(7), 1880–1891 (2019)
5. European Centre for Modern Languages: Día europeo de las lenguas, February 2025. https://edl.ecml.at/Facts/FAQsonsignlanguage/tabid/2741/language/Default.aspx. Accessed 04 Feb 2025
6. Hochreiter, S., Schmidhuber, J.: Long short-term memory. Neural Comput. **9**(8), 1735–1780 (1997)
7. Hoy, M.B.: Alexa, Siri, Cortana, and more: an introduction to voice assistants. Med. Ref. Serv. Q. **37**(1), 81–88 (2018)
8. Kate, J.B.: Smart home system for the deaf and hard-of-hearing (2015). https://api.semanticscholar.org/CorpusID:33454053
9. Kraljević, L., Russo, M., Pauković, M., Šarić, M.: A dynamic gesture recognition interface for smart home control based on Croatian sign language. Appl. Sci. **10**(7), 2300 (2020)
10. Liu, Z., Zhang, C., Peng, H., Xu, Q., Gao, Y.: Drug distribution management system based on IoT. KSII Trans. Internet Inf. Syst. (TIIS) **16**(2), 424–444 (2022)
11. Lupión, M., González-Ruiz, V., Sanjuan, J.F., Medina-Quero, J., Ortigosa, P.M.: Detection of unconsciousness in falls using thermal vision sensors. In: The International Conference on Innovations in Computing Research, pp. 3–12. Springer (2022)
12. Lupión, M., González-Ruiz, V., Sanjuan, J.F., Ortigosa, P.M.: Privacy-aware fall detection and alert management in smart environments using multimodal devices. Internet Things, 101526 (2025)
13. Lupión, M., Medina-Quero, J., Sanjuan, J.F., Ortigosa, P.M.: DOLARS, a distributed on-line activity recognition system by means of heterogeneous sensors in real-life deployments—a case study in the smart lab of the university of Almería. Sensors **21**(2), 405 (2021)
14. Rastgoo, R., Kiani, K., Escalera, S.: Sign language recognition: a deep survey. Expert Syst. Appl. **164**, 113794 (2021)
15. Saunders, B., Camgoz, N.C., Bowden, R.: Continuous 3D multi-channel sign language production via progressive transformers and mixture density networks. Int. J. Comput. Vision **129**(7), 2113–2135 (2021)
16. World Health Organization (WHO): Ear and hearing care, January 2023. https://www.who.int/europe/news-room/questions-and-answers/item/ear-and-hearing-care. Accessed 07 Sept 2024

Automated Detection of Spondylolisthesis in Lumbar X-Rays Using Convolutional Neural Networks

William Moraes da Silva(✉), Silvio César Cazella, and Rafaela Soares Rech

Universidade Federal de Ciências da Saúde de Porto Alegre – UFCSPA, Porto Alegre, RS, Brazil
william.moraes@ufcspa.edu.br

Abstract. Low back pain is a widespread condition and a major contributor to disability worldwide, with spondylolisthesis being one of its associated pathologies. This study evaluates the performance of four convolutional neural network (CNN) architectures—DenseNet201, ResNet152, VGG16, and Xception—in the automatic detection of spondylolisthesis from lumbar X-rays. 10,616 radiographic images from the BUU-LSPINE dataset were used, including anteroposterior (AP) and lateral (LA) projections from 5,308 patients. Three training scenarios were considered: full dataset, LA-only, and AP-only. ResNet152 consistently achieved the highest overall accuracy (88.34%) and AUC (93.64%), while DenseNet201 obtained the highest precision (97.01%) and specificity (98.12%), indicating strong performance in minimizing false positives. VGG16 showed balanced results across metrics, with the highest recall (79.44%) on AP images, whereas Xception underperformed across all evaluation criteria, indicating limited generalization capability. Although the findings are promising, limitations such as a lack of external validation, the absence of advanced preprocessing, and potential demographic bias must be addressed. Future studies should focus on improving generalizability, incorporating explainability tools, and validating models in diverse clinical environments to support reliable deployment in diagnostic workflows.

Keywords: Convolutional Neural Networks · Spondylolisthesis · Medical Image Diagnosis

1 Introduction

Low back pain is a prevalent global condition and a leading cause of work absenteeism and functional disability. Approximately 11.9% of the world's population experiences this issue at some point, with lifetime prevalence reaching up to 84% [1, 2]. In Brazil, prevalence ranges between 59.9% and 62.6%, significantly burdening healthcare systems [3]. Among the associated pathologies, spondylolisthesis is a vertebral displacement condition that may cause pain, instability, and biomechanical alterations [4]. Its incidence varies with age and predisposing factors, influenced by biomechanical stress, repetitive microtraumas, and genetic predisposition [4, 5].

M. Rodrigues et al. (Eds.): ISAmI 2025, LNNS 1776, pp. 62–70, 2026.
https://doi.org/10.1007/978-3-032-14138-5_7

Diagnosis relies on imaging exams, with lumbar spine radiography being a primary method due to its affordability, accessibility, diagnostic accuracy, and minimal radiation exposure [6–9]. However, interpretation requires specialized expertise, posing challenges amid increasing exam demands [9–11]. In this context, artificial intelligence, particularly Convolutional Neural Networks (CNNs), has emerged as a promising tool for diagnosing spinal disorders [12, 13].

CNNs are deep learning architectures capable of automatically extracting features from medical images and identifying pathological patterns [14]. They have shown success in detecting conditions such as pneumonia, breast cancer, and liver diseases [15, 16]. In spinal disorders, CNN-based models demonstrate potential for identifying herniated discs and spondylolisthesis in X-ray and MRI images [17–20].

This study aims to compare transfer learning models applied to CNNs for automatic spondylolisthesis identification in lumbar spine X-rays. By evaluating the performance of pre-trained architectures in extracting biomechanical features, we seek to determine the most effective approaches for diagnosing this condition. This comparative analysis aims to optimize predictive models, enhance clinical decision-making, and reduce specialists' workload.

2 Methodology

The BUU-LSPINE dataset [21], developed through collaboration between Burapha University (BUU) in Thailand and the Korea Institute of Oriental Medicine (KIOM) in South Korea, was utilized in this study with granted permission. It comprises 10,616 X-ray images from 5,308 patients, each with anteroposterior (AP) and lateral (LA) projections. Among them, 1,778 images (889 patients) correspond to individuals with spondylolisthesis, while 8,838 images (4,419 patients) are from healthy individuals. Lateral radiographs were standardized to the left side, and data were anonymized, retaining only sex and age information.

Demographic analysis revealed that spondylolisthesis cases included 616 females and 273 males aged 15–95 years, while healthy cases involved 2,700 females and 1,719 males aged 6–97 years. The images were in ".jpg" format with variable dimensions: AP images ranged from 1,434 × 860 to 3,408 × 3,040 pixels, while LA images ranged from 1,752 × 868 to 3,408 × 3,040 pixels. Despite variations, the dataset was preserved in its original form to maintain analytical integrity.

2.1 Preprocessing and Data Augmentation

To standardize the input for the convolutional neural network, all images were resized to 224 × 224 pixels and converted to grayscale, ensuring a focus on bone structure analysis. Additionally, pixel values were normalized between 0 and 1 to facilitate the convergence of neural network training.

Data augmentation was applied specifically to spondylolisthesis images to enhance the model's generalization ability and address sample size limitations. The applied transformations included rotations of up to 15 degrees, horizontal and vertical shifts of up to 10%, zoom of up to 10%, shear of 0.1, and horizontal flipping, all while preserving the structural integrity of the images. As a result, the dataset was expanded to a total of 15,950 images, consisting of 8,838 healthy cases (55.4%) and 7,112 spondylolisthesis cases (44.6%).

2.2 Pre-trained Architectures

Four pre-trained CNN architectures—DenseNet201, ResNet152, VGG16, and Xception—were used for classification and selected for their high image classification performance [22]. Each network was initialized with ImageNet weights, employing transfer learning. Base convolutional layers remained frozen, while the final convolutional layer was retrained to adapt to dataset-specific features.

2.3 Optimal Number of Neurons

Binary classification experiments were conducted using pre-trained architectures to determine the ideal neuron configuration in the intermediate dense layer. A classification head was added, comprising a GlobalAveragePooling2D layer, a dense layer with varying neuron counts, a dropout layer (0.5), batch normalization for regularization, and an output layer with sigmoid activation.

Models were trained using binary_crossentropy loss, the Adam optimizer (learning rate: 0.0001), and accuracy/loss as evaluation metrics. The training lasted up to 35 epochs with early stopping based on validation loss and ReduceLROnPlateau to prevent overfitting. Seven neuron configurations (32, 64, 128, 256, 512, 1024, 2048) were tested, selecting the best based on accuracy and loss.

2.4 Convolutional Neural Network Models

CNN models were evaluated for spondylolisthesis detection in lumbar X-rays. Analyses were conducted separately for AP and LA projections on the full dataset. Each architecture (DenseNet201, ResNet152, VGG16, Xception) was trained in three models, totaling 12 experiments.

The architectures included a GlobalAveragePooling2D layer, an optimized intermediate dense layer with ReLU activation, and regularization via dropout (0.5) and batch normalization. The output layer contained a single neuron with sigmoid activation for binary classification.

The training lasted up to 35 epochs, with early stopping if validation loss stagnated for three consecutive epochs. ReduceLROnPlateau adjusted the learning rate in response to validation accuracy plateaus. Ten-fold cross-validation ensured robust evaluation, calculating average performance metrics such as accuracy, precision, recall, specificity, F1-score, and area under the ROC curve (AUC).

3 Results

The optimal number of neurons in the intermediate dense layer was determined for each architecture, as shown in Table 1. DenseNet201 performed best with 256 neurons, ResNet152 with 512, VGG16 with 1024, and Xception with 128. Accuracy and loss analysis showed that DenseNet201 achieved 0.8814 accuracy and 0.3163 loss, while ResNet152 reached 0.8759 accuracy and 0.2998 loss. VGG16 obtained 0.8699 accuracy with a 0.3356 loss, whereas Xception had the lowest values, with 0.7588 accuracy and 0.5088 loss.

Table 1. Best results obtained for each architecture.

Architecture	Number of Neurons	Accuracy	Loss
DenseNet201	256	0.8814	0.3163
ResNet152	512	0.8759	0.2998
VGG16	1024	0.8699	0.3356
Xception	128	0.7589	0.5089

After defining the neuron configurations, the compiled models exhibited varying complexity (Table 2). ResNet152 had the highest total parameters (59,422,593), followed by Xception (21,124,393), DenseNet201 (18,815,041), and VGG16 (15,245,121). Regarding trainable parameters, ResNet152 led with 1,050,625, followed by VGG16 (528,385), DenseNet201 (492,545), and Xception (262,657). The high number of non-trainable parameters in ResNet152 highlights its robustness in feature extraction, while Xception's lower trainable parameters correlate with its weaker performance in neuron selection.

Table 2. Parameter distribution of each architecture after adding the optimized dense layer.

Architecture	Total Parameters	Non-Trainable Parameters	Trainable Parameters
DenseNet201	18,815,041	18,322,496	492,545
ResNet152	59,422,593	58,371,968	1,050,625
VGG16	15,245,121	14,716,736	528,385
Xception	21,124,393	20,861,736	262,657

Training the models on the complete dataset, including both lateral (LA) and antero-posterior (AP) projections, revealed ResNet152 as the top-performing architecture in terms of overall accuracy (0.8834). It was followed by DenseNet201 (0.8811), VGG16 (0.8739), and Xception (0.7631). DenseNet201 achieved the highest precision (0.9701), indicating its strong capability to minimize false positives, while ResNet152 also showed high precision (0.9375) and solid recall (0.7911), reflecting its effectiveness in identifying positive cases of spondylolisthesis. DenseNet201 also obtained the highest specificity (0.9812), followed by ResNet152 (0.9575), VGG16 (0.9447), and Xception (0.8595).

Among the architectures, Xception showed the weakest performance, with the lowest F1-score (0.7078) and accuracy, indicating challenges in generalization. AUC values supported these findings: ResNet152 reached the highest score (0.9340), followed by VGG16 (0.9253) and DenseNet201 (0.9064), while Xception had the lowest AUC (0.8248). These results are summarized in Table 3.

Table 3. Average evaluation metrics for models trained on the complete dataset.

Architecture	Accuracy	Recall	Precision	Specificity	F1-Score	AUC
DenseNet201	0.8811	0.7564	0.9701	0.9812	0.8499	0.9064
ResNet152	0.8834	0.7911	0.9375	0.9575	0.8581	0.9340
VGG16	0.8739	0.7859	0.9200	0.9447	0.8475	0.9253
Xception	0.7631	0.6432	0.7886	0.8595	0.7078	0.8248

The same analysis was carried out for images from the lateral (LA) view, as shown in Table 4. ResNet152 achieved the best overall performance, with an accuracy of 0.8809 and a recall of 0.7924, resulting in an F1-score of 0.8555. DenseNet201 also showed strong results, with high precision (0.9500) and recall (0.7624), leading to an F1-score of 0.8456 and the highest specificity among the evaluated architectures (0.9676).

VGG16 exhibited a balanced performance, with a precision of 0.9181, a recall of 0.7880, and a specificity of 0.9434. In contrast, Xception demonstrated the weakest performance, with an accuracy of 0.7646, F1-score of 0.6985, and a specificity of 0.8866, highlighting its difficulty in generalizing to this projection. AUC results mirrored those from the full dataset, with ResNet152 achieving the highest value (0.9319) and Xception the lowest (0.8217).

Table 4. Average evaluation metrics for models trained on LA view images.

Architecture	Accuracy	Recall	Precision	Specificity	F1-Score	AUC
DenseNet201	0.8764	0.7624	0.9500	0.9676	0.8456	0.8978
ResNet152	0.8809	0.7924	0.9296	0.9516	0.8555	0.9319
VGG16	0.8745	0.7880	0.9181	0.9434	0.8479	0.9252
Xception	0.7646	0.6130	0.8138	0.8866	0.6985	0.8217

Finally, the models were also trained using the AP view, as shown in Table 5. The performance pattern was similar to previous results, with ResNet152 achieving the highest accuracy (0.8816), closely followed by DenseNet201 (0.8764). VGG16 recorded the highest recall (0.7944), indicating better detection of positive cases, while DenseNet201 achieved the highest precision (0.9491), suggesting a lower number of false positives. In terms of specificity, DenseNet201 again stood out with the highest value (0.9668), followed by ResNet152 (0.9545) and VGG16 (0.9394).

Xception again showed the weakest overall performance, with an accuracy of 0.7581, F1-score of 0.7079, and the lowest specificity (0.8380), highlighting its difficulty in correctly identifying negative cases. AUC values followed this trend, with ResNet152 achieving the highest (0.9364) and Xception the lowest (0.8175).

Table 5. Average evaluation metrics for models trained on AP view images.

Architecture	Accuracy	Recall	Precision	Specificity	F1-Score	AUC
DenseNet201	0.8764	0.7639	0.9491	0.9668	0.8462	0.9046
ResNet152	0.8816	0.7908	0.9335	0.9545	0.8561	0.9364
VGG16	0.8746	0.7944	0.9135	0.9394	0.8496	0.9295
Xception	0.7581	0.6589	0.7668	0.8380	0.7079	0.8175

A comparative analysis across the three training approaches—using the complete dataset, only lateral (LA) images, and only anteroposterior (AP) images—revealed distinct performance patterns among the evaluated CNN architectures. ResNet152 consistently emerged as the most robust model, achieving the highest accuracy in all scenarios: 0.8834 on the full dataset, 0.8809 on LA images, and 0.8816 on AP images. It also maintained excellent precision and recall balance, with its highest precision observed on the full dataset (0.9701) and solid recall values across projections (0.7911 for the full set, 0.7924 for LA, and 0.7860 for AP). ResNet152 also demonstrated high specificity, reaching 0.9575 on the full dataset and over 0.95 in both views.

DenseNet201 also performed strongly and consistently, excelling in precision (0.9491) and specificity (0.9668) when trained solely on AP images, suggesting enhanced ability to minimize false positives when restricted to a single projection. Its recall values were slightly lower than those of ResNet152 and VGG16, but still competitive (0.7624 on LA and 0.7639 on AP). The model maintained an accuracy of over 0.876 in all training conditions.

VGG16 showed a more balanced behavior, with relatively stable performance across datasets. It recorded the highest recall on the AP projection (0.7944), highlighting its effectiveness in identifying true positive cases, and maintained good precision and specificity, particularly in LA images. Although its accuracy was slightly lower than that of ResNet152 and DenseNet201, it remained above 0.869 in all scenarios.

Xception, however, consistently underperformed. It achieved the lowest accuracy (0.7581 on AP, 0.7646 on LA, and 0.7631 overall), along with the lowest F1-scores and specificity. This indicates clear limitations in generalization and difficulty in correctly identifying both positive and negative cases of spondylolisthesis across projections.

The AUC values corroborated these findings. ResNet152 consistently achieved the highest AUCs, exceeding 0.931 in all scenarios, followed by VGG16 (up to 0.9253) and DenseNet201 (0.9064 on the full set, 0.9226 on AP). Xception again had the weakest AUCs, with the lowest being 0.8175 on AP images.

Regarding image projections, the AP view consistently yielded higher precision and specificity, especially with DenseNet201 and VGG16. This indicates that AP images may present more consistent and distinguishable features for detecting spondylolisthesis in the dataset used, possibly due to clearer visualization of specific vertebral structures.

Comparing these results with previous studies on spondylolisthesis classification, Xu et al. [23] reported a 93.5% accuracy using a two-step model combining YOLOv8 and Res-SE-Net. Varçın et al. [24] achieved 99% accuracy using YOLOv3 and MobileNet. Saravagi et al. [19] reached 100% accuracy by integrating feature selection before classification. Similarly, Güneş et al. [25] analyzed multiple CNN architectures, achieving accuracy levels close to 99%, emphasizing the importance of optimization algorithms. Klinwichit et al. [21] evaluated ResNet152 on a similar dataset, obtaining recall values of 81.31% (AP) and 82.80% (LA), comparable to the current study's results (79.68% and 79.54%, respectively).

These findings reinforce CNNs' potential in supporting clinical diagnosis through lumbar radiograph analysis. When combined with expert evaluation, Zhang et al. [26] demonstrated that deep learning models enhance spinal pathology detection, improve accuracy, reduce errors, and optimize clinical practice decision-making.

4 Limitations

Despite the promising results, certain limitations must be considered. The generalization capability of the models remains a concern, as the BUU-LSPINE dataset [21] may not fully represent the diversity of anatomical patterns and radiographic variations across different populations, imaging protocols, and equipment. Validating these architectures on external datasets from varied clinical settings would improve their robustness and real-world applicability.

The class imbalance between patients with and without spondylolisthesis also posed a challenge. Although data augmentation was applied specifically to increase the number of positive samples, no additional techniques such as class weighting or focal loss were employed. The use of the binary_crossentropy loss function assumes balanced classes and may have limited the model's sensitivity, especially in identifying subtle or atypical cases of spondylolisthesis. Future studies could benefit from loss functions that explicitly address class imbalance to improve recall without compromising specificity.

Furthermore, while the dataset includes age and sex metadata, these variables were not considered in model training or evaluation. As spondylolisthesis prevalence varies with demographic factors, the absence of subgroup analysis prevents the assessment of potential biases. Incorporating demographic attributes into future models may help reveal performance disparities and guide efforts to ensure fairness and equity in clinical applications.

Additionally, extraneous radiographic elements—such as distance markers, positioning labels, and anatomical regions unrelated to the lumbar spine—may interfere with effective feature extraction. Although the dataset was intentionally preserved in its original form to maintain analytical integrity, these elements may introduce noise or spurious correlations. The use of advanced preprocessing techniques, including artifact removal, automatic region-of-interest cropping, or attention-based mechanisms, may enhance the interpretability and precision of CNN-based models in future work.

5 Conclusion

This study evaluated the performance of convolutional neural networks (CNNs) for the detection of spondylolisthesis in lumbar spine radiographs, considering different projections—lateral (LA), anteroposterior (AP), and a combined dataset. Among the evaluated architectures, ResNet152 demonstrated the most consistent and robust performance, achieving the highest overall accuracy (88.34%) and area under the ROC curve (AUC up to 93.64%), indicating reliable identification of positive and negative cases.

DenseNet201 also performed strongly, achieving the highest precision (97.01%) and specificity (98.12%) on the complete dataset, and further excelling in AP projections, suggesting improved reduction of false positives in single-view scenarios. VGG16 maintained balanced results, with the highest recall (79.44%) on AP images, supporting its capability to identify true cases of the condition. In contrast, Xception consistently underperformed, exhibiting the lowest accuracy, recall, and specificity across all datasets, indicating limited generalization capacity.

Despite these promising results, limitations must be considered. The dataset, while comprehensive, may not fully capture anatomical and imaging variations, affecting generalization. Class imbalance remains challenging, potentially influencing the network's ability to recognize pathology-specific patterns. Additionally, extraneous radiograph elements may interfere with feature extraction, impacting model performance.

Future research should focus on validating models with external datasets, refining preprocessing techniques, and integrating hybrid approaches to enhance accuracy and mitigate biases. Importantly, the proposed CNN models do not replace clinical judgment but rather serve as decision-support tools to complement radiological analysis. In particular, such models may be instrumental in identifying spondylolisthesis in high-demand clinical settings, triage workflows, or screening contexts where early and accurate detection can aid timely intervention. When combined with expert interpretation, these tools can enhance diagnostic accuracy, reduce workload, and contribute to more efficient clinical decision-making in spinal pathologies.

Disclosure of Interests. The authors have no competing interests to declare relevant to this article's content.

References

1. Hoy, D., et al.: A systematic review of the global prevalence of low back pain. Arthritis Rheum. **64**, 2028–2037 (2012). https://doi.org/10.1002/art.34347
2. Walker, B.F.: The prevalence of low back pain: a systematic review of the literature from 1966 to 1998. J. Spinal Disord. **13**, 205–217 (2000). https://doi.org/10.1097/00002517-200006000-00003
3. Kaminski, J.S.C., Moreira, A., Mendonça, P.M., Ferreira, L.F.R.: Dados epidemiológicos da dor lombar: prevalência, incidência e incapacidade funcional globalmente e no Brasil. Seven Editora (2023). https://doi.org/10.56238/ciemedsaudetrans-010
4. Martins Filho, D.E., Puertas, E.B., Wajchenberg, M.: Clínica da coluna vertebral. Editora Atheneu (2014)
5. Li, N., Scofield, J., Mangham, P., Cooper, J., Sherman, W., Kaye, A.D.: Spondylolisthesis. Orthop. Rev. **14** (2022). https://doi.org/10.52965/001c.36917

6. Fortes, C.G.M.N.B., Gargioni, L.A., Frast, R.V., Zardo, de Abreu, E., Severo, C.M.: Manifestações clínicas da hérnia discal lombar. Acta Méd. **34** (2013)
7. Sussela, A.O., Bittencourt, A.B., Raymondi, K.G., Tergolina, S.B., Ziegler, M.S.: Hérnia de disco: epidemiologia, fisiopatologia, diagnóstico e tratamento. Acta Méd. **38** (2017)
8. Vialle, L.R., Vialle, E.N., Henao, J.E.S., Giraldo, G.: Hérnia discal lombar. Rev. Bras. Ortop. **45**, 17–22 (2010). https://doi.org/10.1590/S0102-36162010000100004
9. Wang, S., Summers, R.M.: Machine learning and radiology. Med. Image Anal. **16**, 933–951 (2012). https://doi.org/10.1016/j.media.2012.02.005
10. Jha, S., Topol, E.J.: Adapting to artificial intelligence: radiologists and pathologists as information specialists. JAMA **316**, 2353 (2016). https://doi.org/10.1001/jama.2016.17438
11. Tsuneki, M.: Deep learning models in medical image analysis. J. Oral Biosci. **64**, 312–320 (2022). https://doi.org/10.1016/j.job.2022.03.003
12. Alzubaidi, L., et al.: Review of deep learning: concepts, CNN architectures, challenges, applications, future directions. J. Big Data **8**, 53 (2021). https://doi.org/10.1186/s40537-021-00444-8
13. Shen, D., Wu, G., Suk, H.-I.: Deep learning in medical image analysis. Annu. Rev. Biomed. Eng. **19**, 221–248 (2017). https://doi.org/10.1146/annurev-bioeng-071516-044442
14. O'Shea, K., Nash, R.: An introduction to convolutional neural networks. https://arxiv.org/abs/1511.08458 (2015). https://doi.org/10.48550/ARXIV.1511.08458
15. Rajpurkar, P., et al.: CheXNet: radiologist-level pneumonia detection on chest X-rays with deep learning, https://arxiv.org/abs/1711.05225 (2017). https://doi.org/10.48550/ARXIV.1711.05225
16. Seera, M., Lim, C.P.: A hybrid intelligent system for medical data classification. Expert Syst. Appl. **41**, 2239–2249 (2014). https://doi.org/10.1016/j.eswa.2013.09.022
17. Alafeef, M., Fraiwan, M., Alkhalaf, H., Audat, Z.: Shannon entropy and fuzzy C-means weighting for AI-based diagnosis of vertebral column diseases. J Ambient Intell. Hum. Comput. **11**, 2557–2566 (2020). https://doi.org/10.1007/s12652-019-01312-3
18. Fraiwan, M., Audat, Z., Fraiwan, L., Manasreh, T.: Using deep transfer learning to detect scoliosis and spondylolisthesis from X-ray images. PLoS ONE **17**, e0267851 (2022). https://doi.org/10.1371/journal.pone.0267851
19. Saravagi, D., Agrawal, S., Saravagi, M., Chatterjee, J.M., Agarwal, M.: Diagnosis of lumbar spondylolisthesis using optimized pretrained CNN models. Comput. Intell. Neurosci. **2022**, 1–12 (2022). https://doi.org/10.1155/2022/7459260
20. Trinh, G.M., et al.: Detection of lumbar spondylolisthesis from X-ray images using deep learning network. JCM **11**, 5450 (2022). https://doi.org/10.3390/jcm11185450
21. Klinwichit, P., Yookwan, W., Limchareon, S., Chinnasarn, K., Jang, J.-S., Onuean, A.: BUU-LSPINE: a Thai open lumbar spine dataset for spondylolisthesis detection. Appl. Sci. **13**, 8646 (2023). https://doi.org/10.3390/app13158646
22. KERAS: Keras documentation: Keras Applications. https://keras.io/api/applications/. Accessed 02 Nov 2024
23. Xu, Y., et al.: Deep learning model for grading and localization of lumbar disc herniation on magnetic resonance imaging. Magn. Reson. Imaging **61**, 364–375 (2025). https://doi.org/10.1002/jmri.29403
24. Varçın, F., Erbay, H., Çetin, E., Çetin, İ, Kültür, T.: End-to-end computerized diagnosis of spondylolisthesis using only lumbar X-rays. J. Digit. Imaging **34**, 85–95 (2021). https://doi.org/10.1007/s10278-020-00402-5
25. Güneş, H., Hark, C., Akkaya, A.E.: Comparison of deep learning models and optimization algorithms in the detection of scoliosis and spondylolisthesis from X-ray images. Sakarya Univ. J. Sci. **28**, 438–451 (2024). https://doi.org/10.16984/saufenbilder.1246001
26. Zhang, W., et al.: Deep learning-based detection and classification of lumbar disc herniation on magnetic resonance images. JOR Spine **6**, e1276 (2023). https://doi.org/10.1002/jsp2.1276

Augmented Virtuality to Bridge the Gap Between the Real and Virtual Worlds in Neuroarchitecture

José L. Gómez-Sirvent[1,2(✉)], Francisco López de la Rosa[1,3], Roberto Sánchez-Reolid[1,2], José P. Molina[1,2], and Antonio Fernández-Caballero[1,2,4]

[1] Neurocognition and Emotion Unit, Instituto de Investigación en Informática de Albacete (I3A), Universidad de Castilla-La Mancha, Avenida de España s/n, 02071 Albacete, Spain
jose.gomez@uclm.es
[2] Departamento de Sistemas Informáticos, Universidad de Castilla-La Mancha, Avenida de España s/n, 02071 Albacete, Spain
[3] Department of Electrical, Electronic, Automatic and Communications Engineering, Universidad de Castilla-La Mancha, Avenida de España s/n, 02071 Albacete, Spain
[4] Biomedical Research Networking Center in Mental Health (CIBERSAM-ISCIII), 28016 Madrid, Spain

Abstract. Neuroarchitecture is a field that studies how built environments affect human behavior and cognition. Virtual reality (VR) has become a valuable tool for research in this area, as it allows for the simulation and testing of different architectural environments. However, VR systems provide a limited level of immersion, which can affect the realism of the user experience. This paper presents a novel augmented virtuality system that combines chroma-based segmentation with real-time auralization to improve immersion in neuroarchitecture experiments. The main technological challenges associated with this approach are analyzed and the feasibility of the proposed solution is investigated. Performance tests on visual and acoustic latency, along with user feedback, validate the effectiveness of the system. The proposed solution is portable and can be effectively deployed in a variety of environments without the need for specialized infrastructure. This approach has great potential for bridging the gap between the physical and virtual domains, opening up new possibilities for neuroarchitecture and other fields that require high levels of immersion.

Keywords: Neuroarchitecture · Augmented virtuality · Extended reality · Auralization

1 Introduction

The built environment plays a fundamental role in our daily lives, serving as the stage on which people grow, work, interact, and live. The environment influences our routines [3]and our physical and mental well-being [14]. There is a

M. Rodrigues et al. (Eds.): ISAmI 2025, LNNS 1776, pp. 71–80, 2026.
https://doi.org/10.1007/978-3-032-14138-5_8

growing interest in the scientific community to investigate how the characteristics of the built environment affect human well-being. This interest has materialized in the emergence of neuroarchitecture [1], a new interdisciplinary field of study. Neuroarchitecture is considered a component within the broader concept of Ambient Intelligence. By studying how the built environment affects human cognition, behavior, and emotions, neuroarchitecture aims to create personalized spaces that not only adapt to the needs of occupants, but also enhance their well-being. In this context, the use of virtual reality (VR) technologies is becoming increasingly important.

VR has been widely used in recent years to evaluate the impact of various architectural design variables on individuals in a controlled environment. For example, a head-mounted display (HMD) has been used to expose a group of students to classrooms with different lighting levels in order to study how classroom lighting levels affected the memory of the participants [5]. Using similar technology, the impact of room shape on affective states has been measured [15]. HMD technology has also been used effectively in the context of music performance to study the influence of the built environment on musicians [7,8]. This technology makes it possible to immerse individuals in any conceivable architectural environment. However, the virtual experience provided by a conventional HMD is limited to the visual realm, which makes it somewhat restricted by the inability to interact with physical objects or feel the acoustics of the environment. In this context, several extended reality (XR) solutions have been introduced in recent years to soften the boundary between the real and virtual worlds. These range from augmented reality (AR) solutions, where some virtual elements are superimposed on a real environment, to augmented virtuality (AV) solutions, where physical elements are added to a virtual environment [13].

A recent literature review on the use of musical instruments in XR environments [9] highlighted Cave Automatic Virtual Environments (CAVE) and video pass-through HMDs as the key enabling technologies for creating XR audiovisual experiences, with HMDs being the most immersive and cost-effective option. In terms of acoustics, the review highlighted the use of auralization techniques to simulate the acoustics of real spaces in real time from pre-recorded room impulse responses (RIRs). However, it was also noted that previous attempts to use these techniques are limited and have encountered several limitations.

This paper proposes an AV system for neuroarchitectural research using low-cost devices. The main technological challenges are analyzed and the feasibility of the proposed solution is investigated. Our approach aims to bridge the gap between immersive neuroarchitectural research and accessibility by providing a cost-effective alternative without compromising the user experience.

2 Background and Challenges

This section examines the major challenges associated with implementing cost-effective audiovisual environments for neuroarchitectural research. In particular, two critical issues are addressed: enabling users to see their own bodies and interact with physical objects, and designing virtual environments with acoustics that are consistent with the visual experience.

2.1 Virtual Environments

Two virtual environments were modeled to study the feasibility of the proposal: an auditorium and a church. These environments were chosen for their distinctive acoustics, characterized by their large dimensions, geometry and materials, which result in high reverberation times, especially in the case of the church. The environments were modeled using Autodesk Revit 2024 and Unreal Engine 5.3. However, the base geometry of these buildings was obtained from the GRAP database [2], a collection of 35 acoustic models of real buildings. Specifically, the geometries used correspond to the Audimax in Berlin and L'église du Collège Saint-Michel in Fribourg (see Fig. 1).

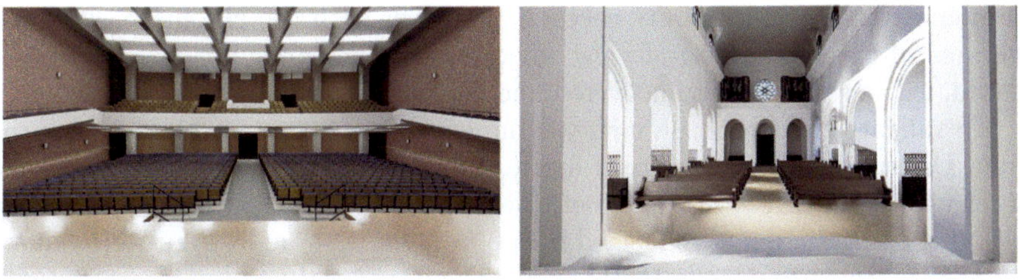

(a) (b)

Fig. 1. Virtual environments used to evaluate the feasibility of the proposed XR system. (a) Audimax auditorium. (b) Church of Collège Saint-Michel.

2.2 Reality-Virtuality Boundary

Video pass-through HMDs offer an affordable and highly immersive alternative to CAVEs, but their use in AV applications is challenging. In order to incorporate physical objects into the virtual environment, or simply to allow the user to see their own body, real-time segmentation of the image captured by the camera in the headset is required, followed by its fusion with the virtual environment. To address this challenge, recent studies have used fixed windows within the virtual environment that allow the user to view parts of the real world environment [4]. In some cases, the presence of these windows is disguised by using a background of the same color as the real and virtual environment [6].

Modern HMDs such as Apple Vision Pro or Meta Quest 3 offer APIs for segmenting objects or the user's hands (https://github.com/oculus-samples/Unreal-OcclusionSample). However, these systems produce relatively inaccurate results because the segmentation masks are updated at a lower frequency than the frame rate of the cameras. Furthermore, these methods require a significant amount of computational power when combining image data and depth sensors. This paper investigates the possibility of replacing these systems with a simple chroma key-based segmentation. Replacing certain colors per alpha channel in an image is a computationally much simpler approach compared to other segmentation systems, allowing for very low latency processing. It can also be used in any pass-through HMD without the need for depth sensors. Although some manufacturers, such as Meta and Apple, block access to the raw camera signal for privacy reasons, they do allow color replacement using look-up tables.

However, chroma key-based segmentation alone is not feasible because the user can look in any direction in 360°. Therefore, to implement the system, it would be necessary to create a uniformly colored sphere in which to place the user. In addition, it would be very difficult to illuminate this setup without real light sources appearing in the virtual environment. To solve this problem, we propose the use of a virtual surface onto which the camera image is projected, which corresponds to the surface of the cyclorama cove used in the physical chroma setup. Thus, it is only necessary to cover with the chroma background the area of the user's field of view where objects or hands are expected to be placed, leaving the ceiling uncovered and facilitating illumination.

To validate this proposal, a simple implementation was performed using the Meta Quest 3 HMD and a green fabric. Figure 2(a) shows the surface used to project the camera feed into the user's field of view. Since the test was performed with the user seated, it was decided to cover only a 180° angle. Figure 2(b) shows the user's view through the HMD, where the user's real hand is integrated into the virtual auditorium environment.

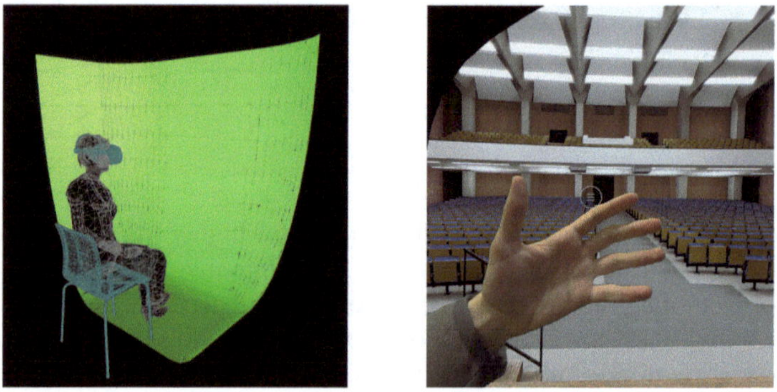

(a) (b)

Fig. 2. Proposed segmentation based on chroma key. (a) Virtual cyclorama cove used for projecting camera images into the virtual environment. (b) Capture of the user's HMD view with the real user hand integrated into the virtual scene.

2.3 Latency in Audio Processing

Simulating the acoustics of a room is relatively straightforward. It simply requires convolving an audio signal with the pre-recorded RIR of the room to reproduce the sound the consumer would experience in that room. Popular video game engines such as Unity or Unreal Engine include plug-ins that allow this to be done easily. However, processing the audio fast enough for the listener to be unaware of the delay is a significant challenge. Humans are able to detect audio latencies of about 10 ms [11], which is little more than the time it takes an audio interface to capture the sound. When a person speaks, the audio heard is a combination of the direct sound coming from the vocal cords into the ears, along with the reverberation caused by the sound bouncing off the various surfaces in the room -early reflections and late reverberations (see Fig. 3).

If a person's voice were auralized with an RIR, they would hear their voice twice because the direct sound reaches the ear much faster than it is possible to capture and process the audio. However, by using a setup that does not block the direct sound, such as a set of loudspeakers arranged in a circular pattern around the user or open-backed headphones, it is possible to present only the reverberation of the environment to the listener. This approach takes advantage of the time delay between the arrival of direct sound and the arrival of the first reflections. A similar method has already been successfully used [10]. However, that proposal generated reverberation parametrically for simple rooms, limiting its ability to simulate complex buildings.

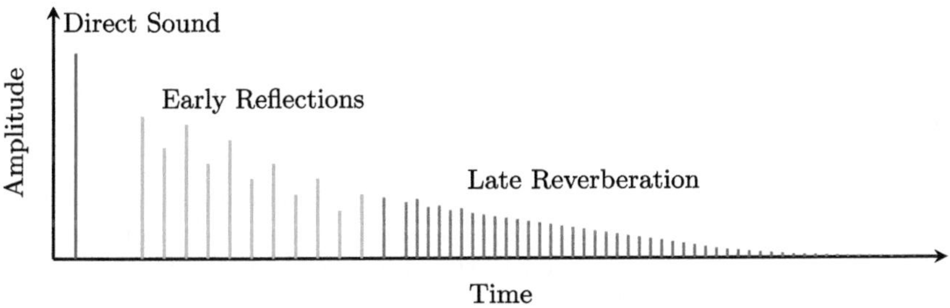

Fig. 3. Example of an impulse response, showing the three main components: direct sound (in red), early reflections (in green), and late reverberation (in blue). (Color figure online)

3 Proposed Augmented Virtuality System

Figure 4 shows a block diagram of the proposed solution. Starting with a basic 3D model of a building, a more visually detailed version is created using Autodesk Revit's building information modeling (BIM) software, which is exported to Unreal Engine for rendering. Meanwhile, the basic 3D geometry is used to calculate the building's RIR using RAVEN, a simulation tool within the Virtual Acoustics (VA) environment (https://www.virtualacoustics.org).

Although the complex geometry modeled in Revit could have been used for this simulation, the use of overly complex geometry in acoustic simulations can lead to implausible solutions. Therefore, simplified geometry is often preferred. In addition, the 3D models obtained from the GRAP database contain predefined acoustic material properties that are lost when exported to Revit. In our implementation, the calculated RIRs are convolved with head-related transfer functions (HRTFs) to generate binaural room impulse responses (BRIRs), allowing users to perceive spatialized audio through headphones.

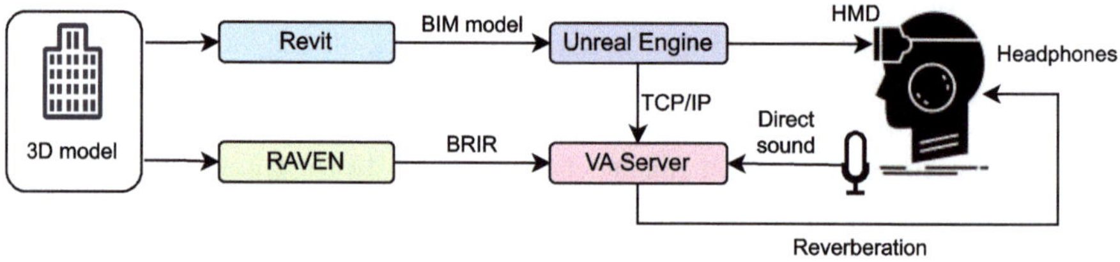

Fig. 4. Block diagram of the proposed system. A BIM model is displayed in the HMD. Head rotation is sent via TCP/IP to the VA server, which uses input from a microphone to generate real-time reverberation.

The VA software is responsible for sound processing by convolving the incoming microphone audio with the pre-computed BRIRs. To compensate for the latency introduced by the audio capture and processing, the BRIRs were trimmed by removing the first 8 milliseconds before the earliest reflections. Different BRIRs were calculated for different head orientations at 5° intervals. Using the TCP/IP protocol, the HMD orientation was sent to the VA server to dynamically assign the closest BRIR. As a result, the listener perceives a spatialized and immersive auditory experience in which the sound adapts to the movements of the head, enhancing the acoustic realism of the virtual environment.

The following hardware was used in the system implementation:

- Meta Quest 3 HMD
- Windows laptop with 32 GB RAM and Nvidia RTX 2070 GPU
- RØDE NT1-A microphone
- AKG K702 headphones
- Presonus Studio 24c audio interface

Although the mentioned microphone and headphones are high-end devices, the system is not dependent on their specific characteristics. More affordable alternatives, such as virtually any wired microphone and headset, could be used. The most expensive component of the system is the laptop. For this study, a PC with 32 GB of RAM and an Nvidia RTX 2070 GPU (released in 2018) was used. While this configuration ensures smooth performance and low latency, it is not strictly necessary to use the same model. Any modern gaming PC, whether desktop or laptop, with equivalent or higher GPU specifications and at least 16

GB of RAM should be able to effectively run the system. CPU usage remained relatively low during the simulation, suggesting that processor performance is not a critical factor. Therefore, it is likely that the system could also run on lower-end machines, though latency would need to be tested and verified in each case.

4 System Evaluation

The evaluation of the developed AV system was based on latency tests and feedback from the researchers about their experience with the solution. Visual latency was monitored using the Meta Quest Developer Hub. Acoustic latency was determined by audio interface configuration parameters such as block size and sample rate. If the sound could not be processed within the time frame defined by these parameters, the sound would be clipped and an alert would be triggered on the VA server. For this reason, iterative testing with different configurations was performed to find the setup that achieved the lowest latency while still ensuring proper audio processing. While the audio interface configuration parameters define the theoretical latency, they do not take into account all contributing factors. For this reason, round trip latency (RTL) was measured for each parameter configuration using RTL Utility (https://oblique-audio.com/rtl-utility.php), which provides an accurate measure of total latency.

In terms of visual performance, all rendering was done on the laptop connected to the HMD via a link cable. The performance of the application was monitored at three different frame rates (72, 90 120 Hz) over a period of 5 min. It was observed that at 72 and 90 Hz, a stable frame rate was maintained in both virtual environments tested. However, 120 Hz, frame rate dropped by 40 Hz during rapid head movements. However, these drops were not perceived by the participant.

With respect to acoustic performance, an RTL latency of about 19 ms was measured. Considering that the impulse responses were trimmed by 8 ms, this leaves an acoustic latency of 11 ms, which would be at the limit of acceptability. In view of this result, the tests were repeated using two laptops connected via TCP/IP. One was responsible for running the visual application in Unreal Engine, while the other was dedicated to the acoustic part running the VA server. In this test, no changes were observed in the frame rate of the HMD. However, the acoustic latency was reduced to 9 ms, and after taking into account the latency compensation in the BRIRs, the effective latency was reduced to 1 ms, which is considered completely undetectable for humans.

On the other hand, the overall user experience with the proposed AV system met expectations, but some limitations and areas for improvement were identified. The proposed segmentation method was quite accurate and computationally efficient, as reflected in the latency tests. However, some shadows and artifacts were observed (see Fig 2(b)). These imperfections can be attributed to irregularities in the background color caused by wrinkles in the fabric or uneven overhead lighting. Therefore, one would expect these artifacts to be minimized in a more controlled setup. Using a background of a rigid material or ironing and tightening the fabric to avoid wrinkles could go a long way toward reducing the shadows observed.

However, the main problem observed was reflections on shiny surfaces. It was observed that the user's watch reflected the green background, causing it to partially disappear from the user's view. Moreover, the proposed solution presents a limitation not found in more complex systems based on depth sensors, since it is not possible to use objects of the same color as the background, nor bright objects that reflect the background color. In addition, outside the area covered by the cylorama surface, the system stops performing segmentation and becomes a conventional VR system.

In terms of sound, the virtual reverberation was tested by speaking into the microphone and playing music through a loudspeaker placed next to the microphone. In both cases, the experience was described as very realistic, but some issues were noted. Loud sounds, such as a scream, or turning up the volume on the speaker too high caused the sound to clipping. Therefore, in order to use this system in future experiments, it would be necessary to limit the sound intensity to avoid clipping. In addition, it was found that the HMD attachment slightly interfered with the support of the headphones on the ear. Therefore, it would be advisable to replace them with in-ear headphones for future experiments.

It should be noted that the user experience evaluation was conducted informally by researchers with extensive experience using VR devices. The user assessment consisted of noting strengths and weaknesses observed in interaction with the system without a set time limit. Due to the absence of standardized protocols and a more diverse sample, this process does not constitute formal validation of the system. Rather, it is a preliminary test intended to identify potential problems and guide future improvements for a more comprehensive usability study.

In summary, despite the limitations encountered, the tests performed provide a promising indication of the feasibility of the proposed AV system. To the best of our knowledge, the combination of real-time segmentation and auralization has never been used before. The closest example to our work may be a virtual recording studio [4], which combined real-time auralization with a window to the real world within the virtual environment, albeit without segmentation. In their experiment, an acoustic latency of approximately 19 ms was measured. This was mainly due to the use of streaming audio for collaborative music creation, which negatively affected the user experience. In contrast, our system achieves effective latency close to zero, so these issues should not arise in this case.

In the field of neuroarchitecture, VR technologies have enabled significant advances, making it possible to conduct numerous experiments without the need to build physical facilities [12]. However, the level of immersion in these experiences, and thus their similarity to real-world environments, is limited. Our system demonstrates that it is possible to significantly increase the level of immersion with relatively inexpensive equipment, paving the way for a more accessible and effective integration of augmented and virtual reality technologies into neuroarchitectural research.

One of the main advantages of the proposed XR solution is its versatility and potential to be integrated as a direct replacement for VR systems used in previous studies. Instead of designing new case studies for this system, most existing VR-based experimental protocols can be adapted to utilize the features of our system. This could be especially beneficial in the field of neuroarchitecture applied to the design of musical spaces [7,8], as the proposed system would allow musicians to see their hands and instruments while enabling the study of room acoustics. This increases experimental possibilities and improves the validity of results.

5 Conclusions and Future Work

This paper has proposed an AV system that can be used to conduct experiments in the field of neuroarchitecture using affordable equipment. The tests conducted demonstrate the effectiveness of the proposed system in providing an immersive experience. Although the system has been conceived for research on the impact of the built environment on individuals, it could be used in many other domains, from video games to training simulators.

Our implementation has demonstrated that immersive AV experiences can be realized in any space without relying on dedicated installations such as a CAVE. This research is intended to serve as a first step in the transition from VR to AV and to provide a path for future studies. Further research could focus on conducting a formal user experience evaluation of the proposed system, or applying this methodology to other experiments in the field of neuroarchitecture.

Acknowledgments. Grant PID2023-149753OB-C21 funded by Spanish MICIU/ AEI/ 10.13039/501100011033 / ERDF, EU. Grant BES-2021-097834 funded by MICIU/AEI/ 10.13039/501100011033 and by "ESF Investing in your future". Grant 2023-PRED-21291 funded by Universidad de Castilla-La Mancha and by "ESF Investing in your future". Grant 2022-GRIN-34436 funded by Universidad de Castilla-La Mancha and by "ERDF A way of making Europe". This work was partially supported by Programa Iberoamericano de Ciencia y Tecnología para el Desarrollo (CYTED) (through Red [225RT0169]) and by CIBERSAM, Instituto de Salud Carlos III, MICIU, and co-funded by "ERDF A way to make Europe".

References

1. Abbas, S., Okdeh, N., Roufayel, R., Kovacic, H., Sabatier, J.M., Fajloun, Z., Abi Khattar, Z.: Neuroarchitecture: how the perception of our surroundings impacts the brain. Biology **13**(4), 220 (2024). https://doi.org/10.3390/biology13040220
2. Ackermann, D., et al.: A ground truth on room acoustical analysis and perception (GRAP). Tech. rep., Technische Universität Berlin (2018). https://doi.org/10.14279/DEPOSITONCE-7003
3. Bayar, R., Türkoğlu, H.: The relationship between living environment and daily life routines of older adults. A/Z: ITU Journal of Faculty of Architecture (2021). https://doi.org/10.5505/itujfa.2021.43410

4. Cairns, P., Hunt, A., Johnston, D., Cooper, J., Lee, B., Daffern, H., Kearney, G.: Evaluation of metaverse music performance with BBC Maida Vale Recording Studios. J. Audio Eng. Soc. **71**(6), 313–325 (2023). https://doi.org/10.17743/jaes.2022.0086
5. Castilla, N., Higuera-Trujillo, J.L., Llinares, C.: The effects of illuminance on students' memory. A neuroarchitecture study. Build. Environ, **228**, 109833 (2023). https://doi.org/10.1016/j.buildenv.2022.109833
6. Fanger, Y., Pfeuffer, K., Helmbrecht, U., Alt, F.: PIANX – a platform for piano players to alleviate music performance anxiety using mixed reality. In: 19th International Conference on Mobile and Ubiquitous Multimedia, pp. 267–276. ACM (2020). https://doi.org/10.1145/3428361.3428394
7. Gómez-Sirvent, J.L., Fernández-Sotos, A., Fernández-Caballero, A., Fernández-Sotos, D.: Assessment of music performance anxiety in a virtual auditorium through the study of ambient lighting and audience distance. Virtual Reality **28**(2) (2024). https://doi.org/10.1007/s10055-024-00947-8
8. Gómez Sirvent, J.L., Fernández-Sotos, D., Sánchez-Reolid, R., de la Rosa López, F., Fernández-Sotos, A., Fernández-Caballero, A.: Pre-occupancy evaluation of a virtual music school classroom: influence of color and type of lighting on music performers. Build. Environ. **246**, 110989 (2023). https://doi.org/10.1016/j.buildenv.2023.110989
9. Gómez-Sirvent, J.L., López de la Rosa, F., Sánchez-Reolid, R., Morales Herrera, R., Fernández-Caballero, A.: Musical instruments in extended reality: A systematic review. International Journal of Human-Computer Interaction, pp. 1–20 (2024). https://doi.org/10.1080/10447318.2024.2431352
10. Klanjscek, N., David, L., Frank, M.: Evaluation of an e-learning tool for augmented acoustics in music education. Music Sci. **4** (2021). https://doi.org/10.1177/20592043211037511
11. McPherson, A., Jack, R., Moro, G.: Action-sound latency: Are our tools fast enough? In: 16th International Conference on New Interfaces for Musical Expression, pp. 20–25 (2016). https://doi.org/10.5281/ZENODO.3964611
12. Rad, P.N., Behzadi, F., Yazdanfar, S.A., Ghamari, H., Zabeh, E., Lashgari, R.: Exploring methodological approaches of experimental studies in the field of neuroarchitecture: a systematic review. Health Environ. Res. Des. J. **16**(2), 284–309 (2023). https://doi.org/10.1177/19375867221133135
13. Rauschnabel, P.A., Felix, R., Hinsch, C., Shahab, H., Alt, F.: What is XR? towards a framework for augmented and virtual reality. Comput. Hum. Behav. **133**, 107289 (2022). https://doi.org/10.1016/j.chb.2022.107289
14. Renalds, A., Smith, T.H., Hale, P.J.: A systematic review of built environment and health. Family Commun. Health **33**(1), 68–78 (2010). https://doi.org/10.1097/fch.0b013e3181c4e2e5
15. Strachan-Regan, K., Baumann, O.: The impact of room shape on affective states, heartrate, and creative output. Heliyon **10**(6), e28340 (2024). https://doi.org/10.1016/j.heliyon.2024.e28340

Personalized Health Assistant: Machine Learning Diabetes Monitoring Integrating IoT and Chatbots

Diogo Ribeiro, Celestino Gonçalves(✉), and Clara Silveira

Instituto Politécnico da Guarda, Guarda, Portugal
1707693@sal.ipg.pt, {celestin,mclara}@ipg.pt

Abstract. This paper presents an integrated system, called Personalized Health Assistant (PHA), that utilizes IoT devices and machine learning to monitor diabetes and promote preventive healthcare. For diabetic patients the IoT devices measure glucose levels, heart rate, SpO_2, and physical activity, while non-diabetic patients had their parameters evaluated based on heart rate and stress levels. Models such as Long Short-Term Memory (LSTM) and Multiple Back-Propagation are used to predict continuous variables, while a Convolutional Neural Network is used to analyze dietary images. Additionally, the system integrates a functional chatbot intended to provide efficient health monitoring and decision support for medical professionals. Among the machine learning models explored, the LSTM network achieved highly satisfactory performance in predicting continuous glucose levels. The model reached a coefficient of determination (R^2) of 0.9875, with a Root Mean Squared Error of 0.0235, and an accuracy of 94.56% in classifying glucose into clinical categories such as hypoglycemia and hyperglycemia. These results highlight the model's effectiveness in both regression and classification tasks within a continuous monitoring context.

Keywords: Diabetes Monitoring · Machine Learning · Deep Learning

1 Introduction

Diabetes mellitus is one of the most prevalent chronic diseases worldwide, causing severe complications that affect quality of life and place a significant burden on healthcare systems [1]. The advancement of technologies such as Internet of Things (IoT) devices, machine learning algorithms, and chatbots enables the creation of innovative solutions for diabetes monitoring and management, providing patients with a more personalized and proactive approach to health control. At the beginning of the process, a detailed analysis is conducted in the hospital environment, where initial clinical values such as blood glucose, blood pressure, and other vital parameters are collected. These values are intended to be entered into the system through a chatbot. Additionally, initial data is analyzed using the K-Means algorithm, with clustering adjustments made to handle high data dimensionality. Long Short-Term Memory (LSTM) models were integrated into the system to predict critical glucose fluctuations. For non-diabetic patients the

M. Rodrigues et al. (Eds.): ISAmI 2025, LNNS 1776, pp. 81–90, 2026.
https://doi.org/10.1007/978-3-032-14138-5_9

Multiple Back-Propagation (MBP) model is applied to predict variables such as glucose and blood pressure, leveraging parallel processing for greater efficiency. A Convolutional Neural Network (CNN) was also applied for dietary analysis. All data is encrypted and securely stored, ensuring a less intrusive medical monitoring approach.

This paper is structured as follows: Sect. 2 covers relevant studies and a comparative analysis; Sect. 3 describes the adopted methodology; Sects. 4 and 5 detail the design, implementation and evaluation of the system; Sect. 6 discusses the results obtained; and finally, Sect. 7 presents the main conclusions.

2 State of the Art

The use of IoT devices and artificial intelligence algorithms has transformed healthcare monitoring, offering advanced solutions for chronic diseases such as diabetes. These advances enable real-time data collection and analysis, providing valuable insights for early risk factor detection and predicting critical metabolic events. Recent studies demonstrate that deep learning models, such as LSTM and CNN, are highly effective in predicting glycemic events and identifying metabolic risk factors. Studies on the genetic epidemiology of type 2 diabetes highlight how hereditary and environmental factors interact, offering opportunities for personalized prevention and new treatment strategies [2]. The interrelationship between metabolic syndrome, diabetes, and cardiovascular risk has been explored, emphasizing the importance of the management of these factors to reduce complications and improve clinical outcomes [3].

LSTM-based temporal models have demonstrated greater accuracy in diabetes-related predictions by using longitudinal blood glucose data. This approach has outperformed traditional methods in recent studies [1]. Another study used deep learning models to predict type 2 diabetes with over 91% accuracy by integrating blood glucose data and other risk factors over time. The application demonstrated reduced predictive errors and advancements in personalized medicine [4]. In the study [5], IoT devices and lightweight algorithms were used to predict blood glucose levels with high real-time accuracy, highlighting the feasibility of cost-effective solutions for continuous monitoring. CNN-based models have been successfully used for diabetes prediction in specific populations, such as industrial workers, enabling risk identification in highly vulnerable populations [6]. CNN were also used to identify dietary patterns associated with glycemic spikes, providing an additional tool for personalized metabolic risk analysis [7].

In [8] an educational chatbot demonstrated effectiveness in improving patient engagement and reducing Hemoglobin A1C (HbA1c) levels, suggesting a promising role in diabetes management. Intelligent chatbots, by interacting directly with patients, help educate them on diabetes management, offering real-time tips and information, thus enhancing self-management of the condition [8, 9].

Based on the highlighted studies, the integration of IoT devices with artificial intelligence algorithms demonstrates an effective approach to continuous monitoring and personalized follow-up for diabetes patients. In this context, some relevant studies from the literature were selected for comparative analysis, considering their contributions and limitations. The studies used reveals that the proposed monitoring system goes beyond the limitations observed in the analyzed studies by integrating different components. This integration results in a robust system capable of addressing the demands of continuous monitoring, glucose prediction, and personalized support. The comparative study (Table 1) illustrates the differences and advancements of the system, Personalized Health Assistant (PHA), compared to the selected studies.

Table 1. Comparison of models and technologies with the proposed solution

Project	Prediction Model	Use of IoT	Glucose Management and Prediction	Chatbot	Computational Efficiency
[7]	LSTM	Yes	Partial Accuracy: 98.26%	No	High
[5]	Random Forest and SVR	Yes	Short-term monitoring	No	High For low-resource devices
[10]	CNN	No	No	No	Moderate
[9]	Not applicable	No	Yes, focused on self-management	Yes	Not applicable
[8]	Not applicable	No	Yes, focused only on self-management	Yes	Not applicable
PHA	K-Means, LSTM, MBP, CNN	Yes	Continuous prediction Accuracy: 94.56%	Yes, focused on personalized alerts	High Support for mobile devices

3 Methodology

The methodology SCRUM [11] was adopted for the development of the project, enabling an agile and iterative approach. The project was structured into sprints with incremental deliveries and continuous validation. In the initial phase, the focus was on creating the visual identity of the application, defining its name, designing the logo, and establishing a consistent branding. The next phases involved the implementation and validation of the machine learning models.

The project utilized an innovative approach for requirements elicitation by leveraging the GPT-4 language model to create interview scripts. The technique was based on the study [12], which proposes the use of prompt chaining to create detailed and realistic interviews for requirements engineering. This method was applied to overcome common challenges in requirements elicitation, such as interview complexity and the difficulty of involving real stakeholders in the early stages of the project. GPT-4 was configured with a specific knowledge base and personalized instructions, enabling the creation of scripts that simulate interactions with stakeholders.

3.1 Functional Requirements

The functional requirements of the PHA system are the following (User Story format):

1. As a patient, I want to view real-time blood glucose levels through my IoT device to monitor my health condition and be alerted about abnormal values. Acceptance Criteria: The system displays updated glucose levels every 5 min; the patient can view detailed graphs of the last 24 h; notifications are sent when levels exceed defined thresholds.
2. As a patient, I want to receive automatic alerts when abnormal glucose values are detected, allowing immediate action. Acceptance Criteria: Real-time notifications provide guidance for actions such as consulting a doctor or taking medication.
3. As a doctor, I want to classify patients into risk categories using the K-Means model to prioritize critical cases and make informed decisions. Acceptance Criteria: The system classifies patients into six risk categories; the algorithm automatically recalculates categories when new data is added.
4. As a patient, I want to monitor glucose, heart rate, blood pressure, and physical activity integrated with my IoT device. Acceptance Criteria: The device collects data continuously and transmits it to the system at regular intervals; the system centralizes information on a dashboard accessible to the doctor.
5. As a patient, I want to interact with a chatbot to report my symptoms and receive guidance based on the provided data. Acceptance Criteria: The chatbot requests additional clinical information in a structured manner; recommendations provided by the chatbot are accurate and tailored to the patient's history.
6. As a doctor, I want detailed reports with graphical data to facilitate clinical evaluation. Acceptance Criteria: Reports include graphs of glucose and other parameters; the system suggests automatic recommendations based on the data.
7. As a patient, I want my data to be stored in an encrypted manner to ensure privacy. Acceptance Criteria: All data is encrypted during storage and transmission; the system complies with General Data Protection Regulation (GDPR).
8. As a patient, I want to input information such as medical history, symptoms, and medication to complement automatic data. Acceptance Criteria: The patient can manually add relevant data; the data is integrated into the system's overall analysis.

3.2 Non-functional Requirements

As non-functional requirements, we can consider the following: (1) Performance - The system must ensure fast and efficient processing to deliver real-time alerts and updates without delays; (2) Security - All sensitive patient data must be protected with robust encryption techniques while ensuring compliance with regulations such as GDPR; (3) Usability - The system interface must be intuitive and accessible, presenting clear messages, interactive graphs, and simplified navigation options to meet the needs of both patients and doctors.

4 Proposed Solution: Personalized Health Assistant

The proposed solution for diabetes monitoring and management integrates advanced technologies and machine learning algorithms to provide real-time monitoring, accurate predictions, and personalized support for patients. The PHA system, whose architecture is illustrated in Fig. 1, is composed of four main components. The first component, not implemented yet, is intended to involve IoT devices, specifically the Dexcom G6 and Fitbit Sense 2, to collect biometric data continuously. The Dexcom G6 to monitor real-time glucose levels, while the Fitbit Sense 2 to record vital parameters such as heart rate, blood pressure, and physical activity [7].

For diabetic patients, glucose levels are measured in real time. In contrast, non-diabetic patients do not have access to continuous glucose monitoring. Instead, glucose levels are predicted using machine learning models based on variables, including heart rate, oxygen saturation, physical activity, and sleep patterns. All data are synchronized in real time with Firebase Firestore, where they are structured and visualized through interactive dashboards, allowing both patients and healthcare professionals to monitor health status continuously and receive timely alerts or recommendations.

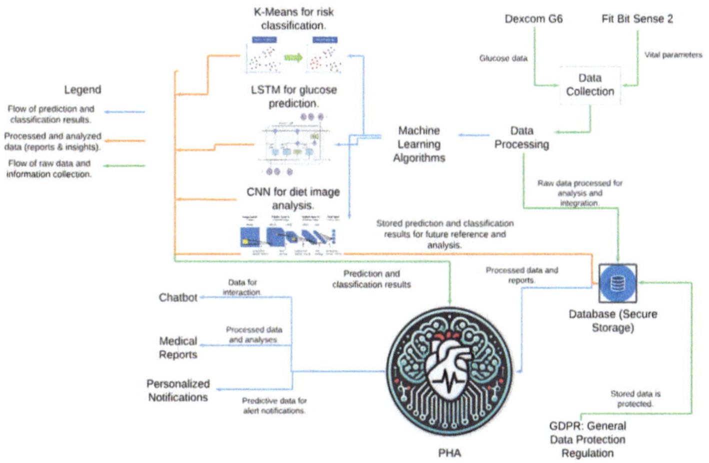

Fig. 1. PHA system architecture integrating IoT, machine learning, and database.

The second component focuses on machine learning models, where various algorithms were implemented for data analysis and prediction. The LSTM model is used to predict fluctuations in glucose levels based on historical data. The MBP algorithm is applied to predict variables in non-diabetic patients, ensuring greater accuracy in complex scenarios. Additionally, the CNN model was trained to analyze images sent by patients, such as diet photos, correlating food intake with metabolic changes. The models were trained using data from the dataset [13], which provides a broad spectrum of information, including glucose measurements, ECG, heart rate, and food images. These data were essential for training the LSTM and CNN models, allowing for more precise and personalized predictions tailored to individual patient needs. Additionally, other datasets were used, incorporating automatically monitored IoT variables for both diabetic and non-diabetic patients: [14–18] and [19].

The third component is the intelligent chatbot, integrated with NLP and LLM, providing direct support to patients through personalized interactions based on collected data and issuing real-time alerts. Currently, the chatbot implementation is still under active development, with critical functionalities progressively being integrated and optimized. Leveraging parallel thread processing, the chatbot will simultaneously monitor multiple patient conditions, ensuring a responsive and dynamic interaction. Additionally, this parallel processing accelerates the collection of supplementary patient data, such as symptoms and dietary habits, optimizing overall system efficiency.

The fourth component focuses on infrastructure and secure data storage, utilizing Firebase to guarantee secure management of sensitive data with end-to-end encryption, fully complying with GDPR regulations. This infrastructure supports rapid data processing necessary for real-time alerts and immediate responses to medical queries. It incorporates essential features such as real-time monitoring, where data continuously collected from IoT devices is presented on intuitive dashboards accessible to both doctors and patients. Additionally, detailed reports and personalized notifications are automatically generated, providing comprehensive support to doctors for informed clinical decision-making. Figure 2 illustrates the PHA system activity diagram.

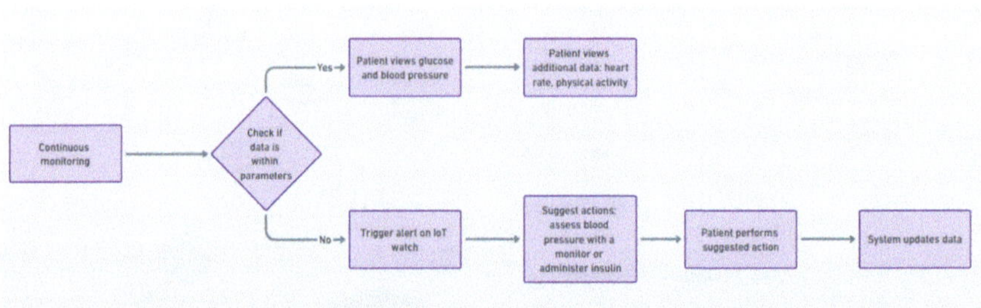

Fig. 2. Activity diagram that will manage alerts and corrective actions.

5 Evaluation

The evaluation section aims to present the results obtained at different stages and approaches of the project. Various methods and algorithms were implemented and tested, ranging from initial patient screening via a chatbot, currently under development, to risk group classification and the application of advanced machine learning models such as LSTM, CNN, and MBP-based methods.

5.1 Grouping Patients into Diabetes Risk Groups (K-Means)

The K-Means model was trained using 1,768 samples from datasets [20] and [21], which were preprocessed to correct missing values and normalize the data. The results proved useful in categorizing patients into six distinct clusters, Fig. 3, considering weighted clinical variables. However, the algorithm has intrinsic limitations, such as the assumption of spherical cluster shapes, which may not accurately reflect the complexity of clinical data.

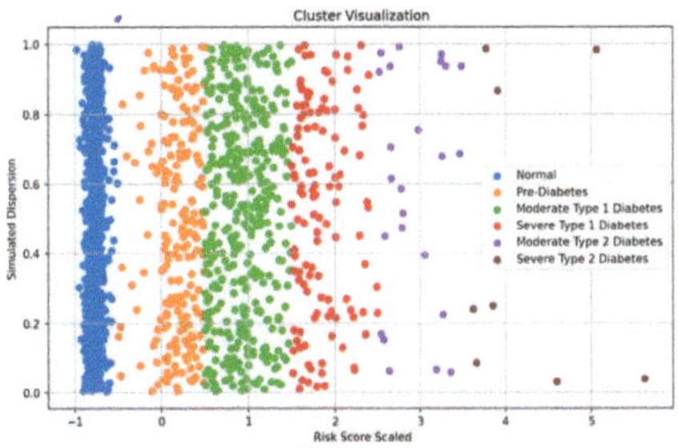

Fig. 3. The clusters categorize patients based on the calculated diabetes risk.

5.2 LSTM Model Evaluation

The LSTM model was developed to estimate blood glucose values based on previous temporal data. It is composed of two stacked LSTM layers with 64 units each, both using ReLU activation, followed by dropout layers with a 30% rate to reduce overfitting, and a final dense layer for regression output. The model was compiled using Adam optimizer, with mean squared error (MSE) as the loss function and mean absolute error (MAE) as the evaluation metric. Training was configured for a maximum of 100 epochs with a batch size of 32, and included an EarlyStopping mechanism with a patience of 10 epochs. This strategy automatically stopped training when validation loss ceased to improve, restoring the best-performing model weights. In practice, the training stopped after 59 epochs. The best model achieved an RMSE of 0.0235, R^2 of 0.9875, and a classification accuracy of 94.56%, when glucose values were categorized into four medical risk levels (hypoglycemia, normal, hyperglycemia mild, and severe), as illustrated in Fig. 4. During validation, scenarios were simulated to detect critical events such as hypoglycemia and high errors in glucose level predictions.

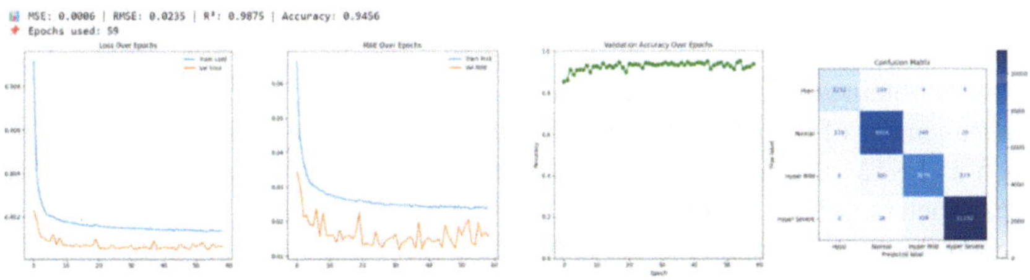

Fig. 4. LSTM model's performance.

During validation, scenarios were simulated to detect critical events such as hypoglycemia and high errors in glucose level predictions. Figure 5 illustrates examples of alerts generated by the LSTM model using validation data, highlighting situations of hypoglycemia, low activity levels, and predictions with significant error relative to the actual measured value. It shows the model's ability to correctly identify risk events and suggest appropriate interventions, such as the recommended administration of insulin.

```
Testing alerts with validation data:
⚠️ ALERT: High prediction error: 0.11. Predicted glucose: 0.13, actual: 0.14
🚨 Hypoglycemia: Glucose 0.13 mg/dL. Administered insulin: 7.00 U.
⚠️ Low activity level: 0.41 g.
⚠️ ALERT: High prediction error: 0.22. Predicted glucose: 0.08, actual: 0.11
🚨 Hypoglycemia: Glucose 0.08 mg/dL. Administered insulin: 4.00 U.
⚠️ ALERT: High prediction error: 0.19. Predicted glucose: 0.60, actual: 0.75
🚨 Hypoglycemia: Glucose 0.60 mg/dL. Administered insulin: 6.00 U.
⚠️ Low activity level: 0.3 g.
```

Fig. 5. Alerts generated by the LSTM model and suggestion of appropriate interventions.

5.3 Evaluation of the CNN and MBP Algorithms

A CNN was implemented to classify the nutritional balance and quality of meals based on food images. The model uses a MobileNetV2 backbone, followed by GlobalAveragePooling, a dense layer with 256 units and ReLU activation, batch normalization, and dropout (40%), ending in a softmax output layer. Training occurred in two phases: a 10-epoch initial phase with frozen layers, followed by fine-tuning up to 150 epochs. The model was trained with batch size 32, using the Adam optimizer and categorical cross-entropy loss, with callbacks for early stopping and learning rate adjustment. Although the training accuracy improved, validation accuracy remained moderate due to class imbalance and the complexity of the multimodal task. Confusion matrices showed a bias toward dominant classes. Future improvements may include the ResNet-50 or EfficientNet, class weighting, data augmentation, stratified splitting and multimodal fusion strategies for textual and nutritional data integration.

Finally, the MBP model was evaluated for continuous variable prediction in other contexts such as stress and blood pressure estimation. It achieved an R^2 of 0.7722 on the Diabetes dataset and above 0.96 in blood pressure prediction using the Sleep Health dataset. These results demonstrate strong performance in time-series prediction, especially when applied to physiological datasets with structured numerical features.

6 Discussion

The results obtained in this study demonstrate significant advances in the monitoring and management of diabetes, aligning with the proposed objectives. The use of models such as LSTM and MBP enabled accurate predictions of blood glucose and other vital parameters, which is consistent with previous studies [7]. However, challenges were encountered when applying the K-Means algorithm for clustering due to the high dimensionality of the data. These challenges underscore the need to improve dimensionality reduction techniques to enhance both accuracy and computational efficiency. The application of parallel processing, including the use of threads and semaphores, was an important differentiator in the system's efficiency. For example, during the initial screening of patients, threads reduced the execution time from 4.52 s (sequential mode) to 1.51 s (parallel mode), optimizing data collection and real-time report generation. Moreover, the use of semaphores to manage the personalized alerts enabled rapid and simultaneous responses in critical situations, such as hypoglycemia, hyperglycemia, tachycardia, and bradycardia. These techniques not only ensured greater efficiency but also demonstrated the system's robustness in environments with multiple simultaneous processes. Although the model's accuracy was high in most analyses, the application of predictive models for more complex data, such as the Physiological Stress Dataset, demonstrated the need for adjustments to handle non-linear data patterns. Additionally, computational efficiency proved to be a positive differentiator of the system, especially on low-power devices such as smartphones and Raspberry Pi, reinforcing its viability for large-scale implementation.

7 Conclusion

This study presented an innovative solution for the monitoring and management of diabetes, combining IoT, machine learning, and chatbots in an integrated system. Furthermore, parallel processing played a crucial role by significantly reducing the execution time of various operations, such as data collection, alert generation, and chart creation, making the system more efficient and scalable.

The main advances include the accurate prediction of physiological variables, computational efficiency on low-power devices, and the application of a less intrusive system for continuous monitoring. Techniques such as the use of threads and semaphores optimized the simultaneous execution of critical processes, such as statistical analysis and risk condition detection, ensuring rapid responses in emergency situations like hypoglycemia and tachycardia.

However, challenges related to the high dimensionality of data and the interpretation of more complex patterns indicate the need to refine the employed techniques and explore new models to handle non-linear patterns. The integration of additional sensors, such as blood pressure monitoring, as well as the implementation of more advanced deep learning techniques, could further improve the system's efficiency and accuracy.

Finally, this work highlights the importance of integrated technological solutions to improve the quality of life for both diabetic and non-diabetic patients, promoting more efficient and personalized monitoring. Future studies may focus on expanding the system's functionality, including large-scale clinical validation, ensuring its applicability in various healthcare contexts.

References

1. Naveed, I., Kaleem, M.F., Keshavjee, K., Guergachi, A.: Artificial intelligence with temporal features outperforms machine learning in predicting diabetes. PLOS Digit. Health **2**(10) (2023)
2. Meigs, J.B.: The genetic epidemiology of type 2 diabetes: opportunities for health translation. Curr. Diabetes Rep. **19**, 1–8 (2019)
3. Mottillo, S.: The metabolic syndrome and cardiovascular risk: a systematic review and meta-analysis. J. Am. Coll. Cardiol. **56**(14), 1113–1132 (2010)
4. Association, A.D.: Diagnosis and classification of diabetes mellitus. Diabetes Care **37**(Suppl. 1), S81–S90 (2014)
5. Rodríguez-Rodríguez, I., Campo-Valera, M., Rodríguez, J.-V., Frisa-Rubio, A.: Constrained IoT-based machine learning for accurate glycemia forecasting in type 1 diabetes patients. Sensors **23**(7), 3665 (2023)
6. Zhao, J., Gao, H., Yang, C., An, T., Kuang, Z., Shi, L.: Attention-Oriented CNN method for type 2 diabetes prediction. Appl. Sci. **14**(10), 3989 (2024)

7. Rajalakshmi, R., Sivakumar, P., Kumari, L.K., Selvi, M.C.: A novel deep learning model for diabetes mellitus prediction in IoT-based healthcare environment with effective feature selection mechanism. J. Supercomput. **80**(1), 271–291 (2024)
8. Nassar, C., Dunlea, R., Montero, A., Tweedt, A., Magee, M.: Feasibility and preliminary behavioral and clinical efficacy of a diabetes education chatbot pilot among adults with type 2 diabetes. J. Diabetes Sci. Technol. **19**(1), 54–62 (2025)
9. Lokman, A.S., Zain, J.M., Komputer, F.S., Perisian, K.: Designing a chatbot for diabetic patients. In: International Conference on Software Engineering & Computer Systems (ICSECS 2009), pp. 19–21 (2009)
10. Wu, J.-H., et al.: Risk prediction of type 2 diabetes in steel workers based on convolutional neural network. Neural Comput. Appl. **32**, 9683–9698 (2020)
11. Sutherland, K., Schwaber, J.: The Scrum guide – the definitive guide to scrum: the rules of the game. Scrum.org (2020)
12. Görer, B., Aydemir, F.B.: GPT-powered elicitation interview script generator for requirements engineering training. In: IEEE 32nd International Requirements Engineering Conference, pp. 372–379 (2024)
13. Dubosson, F., Ranvier, J.-E., Bromuri, S., Calbimonte, J.-P., Ruiz, J., Schumacher, M.: The open D1NAMO dataset: a multi-modal dataset for research on non-invasive type 1 diabetes management. Inform. Med. Unlocked **13**, 92–100 (2018)
14. Cardiovascular disease dataset. https://www.kaggle.com/datasets/sulianova/cardiovascular-disease-dataset. Accessed Jan 2025
15. ECG Heartbeat Categorization Dataset. https://www.kaggle.com/datasets/shayanfazeli/heartbeat. Accessed Jan 2025
16. Human Stress Detection in and through Sleep. https://www.kaggle.com/datasets/laavanya/human-stress-detection-in-and-through-sleep. Accessed Jan 2025
17. Sleep Health and Lifestyle Dataset. https://www.kaggle.com/datasets/uom190346a/sleep-health-and-lifestyle-dataset. Accessed Jan 2025
18. Koldijk, S., Sappelli, M., Verberne, S., Neerincx, M., Kraaij, W.: The swell knowledge work dataset for stress and user modeling. In: Proceedings of the 16th International Conference on Multimodal Interaction, ICMI 2014, pp. 291–298 (2014)
19. Diabetes_Dataset_With_18_Features. https://www.kaggle.com/datasets/pkdarabi/diabetes-dataset-with-18-features. Accessed Jan 2025
20. Pima Indians Diabetes Database. https://www.kaggle.com/datasets/uciml/pimaindians-diabetes-database. Accessed Jan 2025
21. Rashid, A.: Diabetes Dataset. Mendeley Data (2020)

Exploring Time Series xAI Techniques for 5G/6G Networks

Fábio Silva[1(✉)], José Ribeiro[1], Cesar Analide[2], and Ricardo Santos[1]

[1] CIICESI, ESTG, Insituto Politécnico do Porto, Felgueiras, Portugal
{fas,jfr,rjs}@estg.ipp.pt

[2] Department of Informatics, ALGORITMI Center, University of Minho, Braga, Portugal
analide@di.uminho.pt

Abstract. This study explores the applicability of explainable artificial intelligence (xAI) techniques in the analysis of deep learning models for anomaly detection in 5G/6G networks. With the increasing complexity of networks and network traffic, the mission to guarantee the security access points and devices against attacks and intrusions is also larger. Models used for these tasks operate like black boxes, making it difficult to understand and interpret their decisions at a human level. To address this challenge, we devised a case study with a real world dataset and a performant deep learning anomaly detection algorithm and implemented strategies to generate human understandable explanation through xAI algortihms. xAI can provide insights into the factors that lead to the detection of anomalies, allowing for greater transparency and reliability in the process. This work is part of the context of intelligent networks and is aligned with initiatives such as the Privateer project, contributing to the evolution of security in 5G/6G infrastructures. The integration of deep learning and xAI facilitates interaction between human operators and automated systems, promoting greater control over decision-making in modern networks.

Keywords: 5G/6G networks · anomaly Detection · xAI

1 Introduction

Artificial Intelligence (AI) has addressed challenges in 5G and 6G networks, particularly in traffic monitoring and anomaly detection. Deep learning models have shown effectiveness in identifying patterns that may suggest faults, congestion or attacks [7]. However, these models are opaque, making it difficult to understand their decisions [4]. The interpretability of AI models, known as eXplainable Artificial Intelligence (xAI), has been an area of great interest in the scientific and industrial community. The goal of xAI is to increase the transparency of deep learning models, making their predictions understandable to human operators

M. Rodrigues et al. (Eds.): ISAmI 2025, LNNS 1776, pp. 91–100, 2026.
https://doi.org/10.1007/978-3-032-14138-5_10

and network experts [6]. In 5G/6G network security and optimization applications, explainability plays a role in providing justification for anomaly detections and recommendations generated by algorithms. Techniques such as SHAP (SHapley Additive Explanations) [10] and LIME (Local Interpretable Model-agnostic Explanations) [14] have been widely used to provide insights into the contribution of each variable to the final result of the models. There is however, a time dependency that most models for network traffic anomaly detection implement as a timeseries problem which also require personalization of common xAI algorithms.

In this study, we explore the application of interpretability methods to the analysis of traffic anomalies in 5G and future 6G networks. In particular, we investigate how xAI techniques can be used to explain decisions made by deep learning models that detect anomalous patterns in network traffic. The problem addressed involves the analysis of traffic filtered at access points, where it is possible to distinguish between normal behavior and potential intrusion attempts or cyber attacks.

2 State of the Art

Advances in artificial intelligence techniques have led to the development of increasingly complex models, applied in various fields, including intelligent communication networks. In the context of 5G networks and future 6G infrastructures, deep learning models have been used to analyze and monitor traffic, enabling the detection of patterns and possible anomalies. However, the opaque nature of these models represents a significant challenge for their adoption in these scenarios, where transparency of decisions is useful [12,14].

Explainable artificial intelligence (xAI) has emerged as a key field for increasing the interpretability of models, providing the means to understand the reasons behind the predictions made by algorithms. Techniques such as SHAP (SHapley Additive Explanations), a method based on Shapley values from game theory that allows the influence of each variable on the model's final prediction to be quantified, and LIME (Local Interpretable Model-agnostic Explanations) have been widely used to reveal the influence of different variables on the decisions of anomaly detection models. In addition, strategies based on autoencoders and attention mechanisms have been explored to improve model transparency in modern network scenarios.

In this context, the applicability of xAI in 5G/6G networks enables greater control over intelligent infrastructures, allowing effective interaction between human operators and automated systems, promoting security and reliability in decision-making processes.

2.1 Anomaly Dectection in Network Traffic

Detecting anomalies in 5G/6G networks is a current challenge, given the increasing complexity and dynamism of data traffic. Different approaches have been explored for this purpose, including methods based on deep learning, statistical techniques and hybrid models. Deep neural networks, such as autoencoders and attention-based models, have proven effective in identifying anomalous patterns by learning latent representations of traffic data. Models such as LSTM(Long Short-Term Memory), a type of recurrent neural network (RNN) designed to capture long-term temporal dependencies in time series, [5,13] and GRU (Gated Recurrent Unit), a simplified variant of LSTM that also captures temporal patterns in sequential data with less computational complexity, have been widely used to capture temporal dependencies in mobile network scenarios, enabling early detection of attacks such as DDoS and malicious intrusions [3,8].

In addition to deep learning-based techniques, statistical approaches such as threshold rules and clustering algorithms (e.g. DBSCAN and k-means) are often used to identify atypical behavior. However, the opaque nature of deep learning models poses challenges in terms of interpretability. Explainability methods such as SHAP and LIME have been applied to understand the decisions of these algorithms, allowing network operators to identify the most influential factors in detecting anomalies. xAI enhances transparency and monitoring reliability in 6G networks.

2.2 Explicability in Machine Learning Models

The explainability of machine learning models has become useful for ensuring transparency and reliability, especially in sensitive applications such as telecommunications and network security [1]. Explainable artificial intelligence (xAI) seeks to make models more comprehensible to humans, allowing the interpretation of decisions made by "black box" models such as deep neural networks. In the context of traffic analysis in 5G/6G networks, where models need to detect anomalies in continuous data flows, explainability should try to help understand situations.

The interpretation of models applied to time series is still an open challenge. Most xAI techniques were originally developed for conventional supervised models, which requires adaptations when applied to continuous data streams [12]. However, methods such as SHAP and LIME have been used to evaluate the influence of input variables on the predictions of anomaly detection models in networks [10,14]. Other approaches include gradient-based methods such as Integrated Gradients and Layer-wise Relevance Propagation (LRP), which analyze the model's sensitivity to variations in inputs, and visualization techniques such as saliency maps and feature visualization, widely used in convolutional networks but adaptable to temporal networks [15,17]. Although not applied here, these approaches can be integrated into the modular architecture designed to support various xAI techniques.

Applying these techniques to 5G/6G time dependent network traffic anomaly detection models provides a better understanding of the decisions made, providing justification for detecting attacks or anomalous patterns. The use of interpretability techniques in 5G/6G networks is useful due to the complexity of these infrastructures [...] making the decisions of these models more understandable to network operators and experts. 6G implementation proposals, such as in [11], and [9] have explored the integration of xAI techniques into intelligent telecommunications systems, with the aim of improving the transparency of anomaly detection algorithms in high-connectivity environments.

3 Explainable Anomaly Detection in 5G/6G Networks

Growing connectivity and device diversity make securing mobile networks a relevant challenge. The complexity of traffic patterns and the need for real-time monitoring require advanced solutions that combine machine learning and explainable artificial intelligence (xAI) techniques to make the models more understandable and reliable for operators and experts. The architecture proposed, Fig. 1 consists of flexible microservices, containerized, responsible for every step for the production of xAI insights through different techniques and strategies.

Fig. 1. Architecture.

The architecture developed consists of a set of microservices, organized to ensure flexibility and scalability in the explainable analysis of anomalies. The microservices are divided by function: (1) a database management microservice, responsible for managing the data collected from the 5G infrastructure; (2) a model management microservice, responsible for loading the trained models and making them available for analysis; and (3) microservices dedicated to explainability techniques. This structure allows for modular integration, facilitating the future introduction of new detection algorithms or xAI techniques.

3.1 Dataset and Anomaly Detection Models

The test environment consisted of two cells with a total of nine UEs connected to the same core network. The 5G network infrastructure was implemented with the Amarisoft Callbox Mini and Amarisoft Classic solutions, the latter responsible for hosting the core of the 5G network. The set of UEs included Huawei P40 smartphones, Raspberry Pi 4 microcomputers equipped with 5G modules, industrial 5G routers, WiFi-6 mobile hotspots and a CPE Box. During data recording periods, the devices generated network traffic in a programmed way, simulating real communication patterns producing the NCSRD-DS-5GDDoS dataset [16]. This dataset contains detailed radio and core metrics from 5G networks, including sporadic Distributed Denial of Service (DDoS) attacks initiated by malicious users (UEs).

The anomaly detection model used was trained with an autoencoder architecture based on the LSTM deep learning algorithm, using a pre-processed version of the NCSRD-DS-5GDDoS [2] dataset. Training was carried out with benign test data from the dataset, organized in a time series format by device, selecting the 8 most relevant variables in time windows of 120 instants. Each input vector was thus built with 960 values (120×8), representing the temporal context of each device. For interpretability purposes, during the application of the SHAP technique, the vector is again segmented into its original components, making it possible to analyze the influence of each variable over time and identify the instants with the greatest impact on the model's decision. The model identified anomalies with 99% accuracy.

3.2 xAI Strategies for 5G/6G Time Series Environments

This study uses a pre-trained model to apply interpretability techniques. The approach adopted used SHAP exploring the nature of the anomaly detection algorithm and the properties of the timeseries dataset to devise strategies that can help explain the influence of input features accross data windows to the final detection. The model analyzed takes as input a time window of 120 instants, with 8 attributes per instant, resulting in an input vector of dimension 960. Each instance analyzed by XAI corresponds to this structure, showing how temporal variations of features influence the models anomaly detection.

Two main strategies were implemented to explain the predictions. Figure 2 shows two different approaches to explaining the model's decisions. The image on the left represents the "Loss by Attribute" strategy, which evaluates the average impact of each attribute over the time window. The image on the right represents the "Loss per Instance" strategy, which assesses the influence of a specific variable over multiple instances of the time series. Both representations aim to visually illustrate the contribution of temporal data to the model's decision.

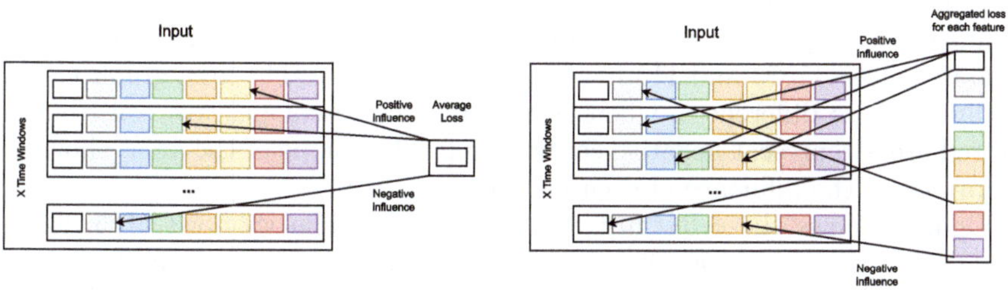

Fig. 2. xAI timeseries strategies.

3.3 Analysis of Influence by Attribute (Loss by Attribute)

In this section, we present the first xAI-based explanation approach, in which we analyze the global contribution of all attributes in anomaly detection. To do this, we use techniques such as SHAP to calculate the influence of each variable over the time sequence and assess its relevance in predicting the model. This approach makes it possible to see which attributes, in general, have the most influence on the model's decision [10]. As a result, it was possible to see that certain attributes, such as cell_x_ul_retx and cell_x_ul_tx, had a significantly greater impact on the classification of anomalies, while other attributes had a lesser influence. In addition, it was observed that the importance of certain variables can vary over time, reflecting changes in network traffic patterns.

These 5G radio-layer metrics help interpret traffic and identify anomalies. The cell_x_ul_retx variable refers to the number of retransmissions on the uplink, which can indicate failures or congestion in data transmission. Cell_x_ul_tx represents the total volume of data sent on the uplink, while cell_x_dl_tx refers to the volume of data transmitted on the downlink, both direct indicators of network activity and performance. The variables dl_total_bytes_non_incr and ul_total_bytes_non_incr indicate the volume of data that has not increased over a period of time, and may be associated with periods of inactive traffic or irregular patterns that suggest anomalous behavior. Explaining these attributes clarifies their influence and justifies the model's output.

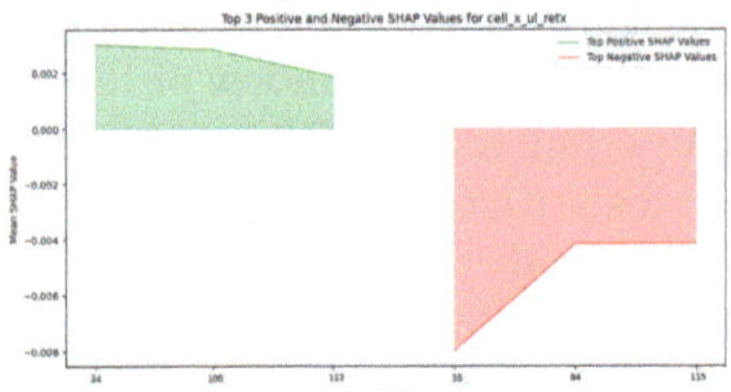

Fig. 3. Top 3 SHAP values for cell_x_ul_retx (Area chart).

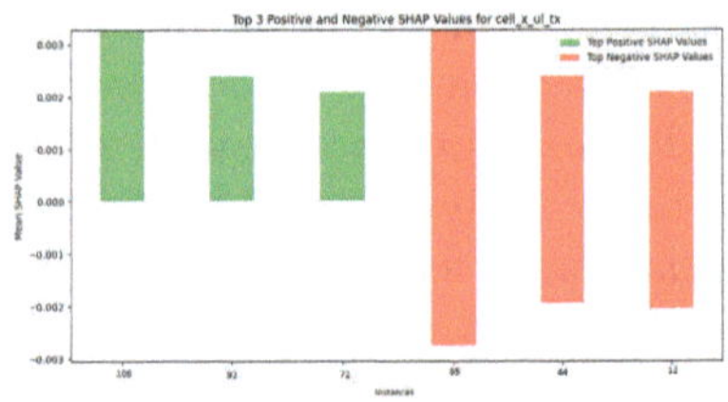

Fig. 4. Top 3 SHAP values for cell_x_ul_tx (Bar chart).

Figures 3 and 4 show the three main attributes with the greatest influence on the model's prediction. These graphs illustrate the positive and negative contribution of the attributes selected on the basis of the SHAP values. Figure 3 shows the cell_x_ul_retx variable in the time windows (24, 100, 112, 35, 84, 115). The instances with a positive impact on the model's prediction are 24, 100 and 112, while instances 35, 84 and 115 have a negative impact. The area graph visually highlights these differences, where green areas represent positive contributions and red areas represent negative contributions. The slope of the areas shows the magnitude of the influence of each instance.

Figure 4 shows the cell_x_ul_tx variable in the time windows (106, 92, 72, 85, 44, 12). Instances 106, 92 and 72 have a positive impact on the model's forecast, while instances 85, 44 and 12 contribute negatively. The height of the bars indicates the magnitude of the influence of each instance, with green bars representing positive SHAP values and red bars representing negative values.

This analysis shows that the variables associated with uplink traffic (ul_tx and ul_retx) significantly influence the model's predictions and are determining factors in detecting network anomalies. This information can be used to improve the interpretation of the model's decisions and refine future analyses.

3.4 Analysis of Influence by Instance (General Loss)

In this section, we present the second explanation approach applied in this study, which focuses on the influence of a specific variable over the different instances analyzed. Instead of considering all attributes simultaneously, this strategy aims to understand how a single attribute impacts the model's decision in different time windows [14].

This analysis makes it possible to check the variability of an attribute's importance at different points in the time series. For example, the variable cell_x_dl_tx was found to have a high impact in certain time windows associated with traffic peaks, indicating a strong correlation with anomalous events.

In this way, the influence-by-attribute approach offers a detailed perspective on the behavior of individual variables, helping to identify specific patterns that may be related to the occurrence of anomalies.

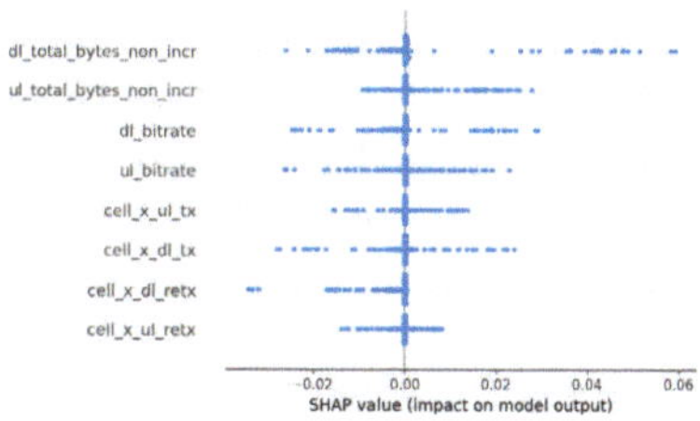

Fig. 5. Scatterplot of SHAP values for different attributes.

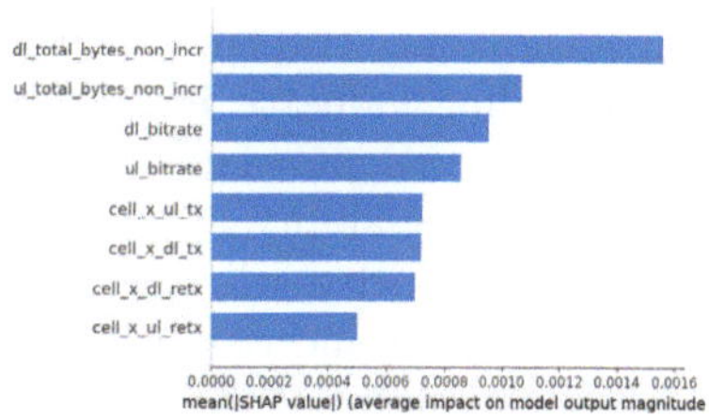

Fig. 6. Average impact of attributes on the model based on SHAP values.

Figures 5 and 6 illustrate the importance of the attributes for predicting the model in instance 0, considering the SHAP values. In Fig. 5, each point represents a SHAP value associated with a specific attribute. The distribution of points along the horizontal axis indicates the relative impact of each variable on the model's decision. It can be seen that attributes such as dl_total_bytes_non_incr and ul_total_bytes_non_incr show greater variability, suggesting that they have more influence in predicting anomalies. In Fig. 6 shows the average absolute value of the SHAP coefficients for each variable, highlighting the attributes with the greatest overall influence. It can be seen that the attributes dl_total_bytes_non_incr and ul_total_bytes_non_incr have the highest impacts, while variables such as cell_x_ul_retx have a lower influence. The analysis highlights how attribute impact varies over time, aiding interpretation and anomaly detection.

3.5 Critical Review

In the first strategy, the analysis of a specific variable throughout the instances revealed interesting patterns. For example, when selecting the variable dl_bitrate, it was observed that its influence varied according to the instance analyzed. In certain instances, this variable had a very high impact on predicting the anomaly, while in others its effect was less significant. This behavior suggests that the relevance of the variables may be sensitive to the context of the instance analyzed, reinforcing the need for interpretive techniques to understand the dynamics of the model.

In the second strategy, it was possible to observe that certain variables had a consistent impact throughout the time sequence. For example, the variable dl_bitrate proved to be one of the most influential in predicting anomalies, indicating that significant variations in the downlink transmission rate may be strongly associated with unusual events in the network. Other variables, such as ul_bitrate and cell_x_dl_retx, also had a significant influence, standing out at specific moments in the time sequence as determining factors for the model's classification.

These interpretability techniques improve transparency, allowing experts to understand and validate automated decisions in 5G/6G networks. The combination of these two strategies allowed for an in-depth analysis of the model's behavior, providing valuable insights into the factors that most influence anomaly detection in 5G/6G networks. The results obtained demonstrate the importance of explainability in machine learning models, especially in scenarios such as network security, where the correct interpretation of predictions can help in decision-making and risk mitigation.

The evaluation of the explanations considered the SHAP values, assuming that values close to zero indicate irrelevance and values close to one reflect high influence, thus allowing an estimate of the fidelity of the explanations generated. Although the detection model used only returns a binary classification (anomaly or normal), the introduction of explainability techniques such as SHAP has made it possible to overcome this limitation, providing a detailed view of the variables that influenced each decision. This capability represents an added value compared to traditional approaches without interpretability, which do not make it possible to understand the factors behind the classifications. Although not compared directly, the architecture enhances model understanding and supports real-world use where transparency matters.

4 Conclusion and Future Work

The implementation of xAI strategies enabled an in-depth analysis of the interpretability of the LSTM model applied to anomaly detection in 5G/6G networks. Two main approaches were applied throughout this study: loss per attribute, which assesses the relative importance of each variable over the time series, and general loss, which measures the relevance of attributes in identifying anomalous patterns.

The first approach identified key variables by analyzing their contribution over the 960-entry window. It was observed that certain attributes had a considerable impact at specific moments in the time series, directly influencing the model's decisions. This understanding is intended to improve the explainability of the model and direct efforts towards improving the quality of the data and pre-processing techniques. In the second approach, the importance of attributes in detecting anomalies was analyzed globally, making it possible to see which variables were most strongly associated with identifying anomalous events. This strategy highlighted the relationship between the influence of certain attributes and the occurrence of unusual patterns, providing valuable insights into the factors that contribute to the classification of an anomaly.

This study opens the way to several possibilities for future work which include analysing the importance of each data windows towards the final outcome. Another aspect for future research is adapting the model to different traffic and attack scenarios in 5G/6G networks. Validating xAI strategies in new contexts could help improve the robustness of the model and its applicability in real systems. This reinforces the importance of interpretable models in security contexts, supporting trust in automated systems and informed decision-making. Future work will evaluate latency, overhead and scalability for deployment in real-time distributed systems.

Acknowledgments. This work was performed in the scope of the Smart Networks and Services Joint Undertaking (SNS JU) under the EU Horizon Europe programme PRIVATEER under Grant Agreement No. 101096110, This work has also been supported by national funds through FCT Fundação para a Ciência e Tecnologia through project UIDB/04728/2020.

References

1. Adadi, A., Berrada, M.: Peeking inside the black-box: a survey on explainable artificial intelligence (XAI). IEEE Access **6**, 52138–52160 (2018). https://doi.org/10.1109/ACCESS.2018.2870052
2. Argyriou, L., Karamatskou, A.: Privateer deliverable d3.1: decentralised robust security analytics enablers - rel.a. (2024). https://doi.org/10.5281/zenodo.12530825
3. Cho, K., Merrienboer, B., Gulcehre, C., Bougares, F., Schwenk, H., Bengio, Y.: Learning phrase representations using RNN encoder-decoder for statistical machine translation (2014). https://doi.org/10.3115/v1/D14-1179
4. Doshi-Velez, F., Kim, B.: Towards a rigorous science of interpretable machine learning. arXiv, Machine Learning (2017)
5. Greff, K., Srivastava, R.K., Koutnik, J., Steunebrink, B.R., Schmidhuber, J.: LSTM: a search space odyssey. IEEE Trans. Neural Networks Learn. Syst. **28**, 2222–2232 (2017). https://doi.org/10.1109/TNNLS.2016.2582924
6. Guidotti, R., Monreale, A., Ruggieri, S., Turini, F., Giannotti, F., Pedreschi, D.: A survey of methods for explaining black box models. ACM Comput. Surv. **51** (9 2018). https://doi.org/10.1145/3236009

7. Heaton, J.: Ian goodfellow, yoshua bengio, and aaron courville: deep learning. Genet. Program. Evolvable Mach. **19**, 305–307 (2017). https://doi.org/10.1007/S10710-017-9314-Z
8. Hochreiter, S., Schmidhuber, J.: Long short-term memory. Neural Comput. **9**, 1735–1780 (1997). https://doi.org/10.1162/NECO.1997.9.8.1735
9. Liyanage, M., et al.: Advancing security for 6G smart networks and services 1169–1174 (2024). https://doi.org/10.1109/EuCNC/6GSummit60053.2024.10597131
10. Lundberg, S.M., Lee, S.I.: A unified approach to interpreting model predictions. In: Proceedings of the 31st International Conference on Neural Information Processing Systems, pp. 4768–4777, NIPS 2017. Curran Associates Inc., Red Hook, NY, USA (2017)
11. Masouros, D., et al.: Towards privacy-first security enablers for 6G networks: the privateer approach, pp. 379–391 (2023). https://doi.org/10.1007/978-3-031-46077-7_25
12. Molnar, C.: Interpretable Machine Learning. Leanpub (2020)
13. Olah, C.: Understanding LSTM networks (2015). https://research.google/pubs/understanding-lstm-networks/
14. Ribeiro, M.T., Singh, S., Guestrin, C.: "why should i trust you?" explaining the predictions of any classifier. In: Proceedings of the ACM SIGKDD International Conference on Knowledge Discovery and Data Mining ,13–17-August-2016, pp. 1135–1144 (2016). https://doi.org/10.1145/2939672.2939778
15. Samek, W., Wiegand, T., Müller, K.R.: Explainable artificial intelligence: understanding, visualizing and interpreting deep learning models. arXiv:1708.08296 (2017)
16. Hellas, S.: NCSRD-DS-5GDDoS: 5G radio and core metrics containing sporadic DDoS attacks. Nat. Centre Sci. Res., Dataset. (2024) https://doi.org/10.5281/ZENODO.10671494, https://zenodo.org/records/10671494
17. Sundararajan, M., Taly, A., Yan, Q.: Axiomatic attribution for deep networks. In: Proceedings of the 34th International Conference on Machine Learning - Volume 70, pp. 3319–3328, ICML 2017. JMLR.org (2017)

Ultrasound-Based Detection of Pediatric Appendicitis Using ViT

Carlos Luis Fernández-Santana[1(✉)], Pablo Negre[1], Pablo Enrique Guillem[2], and Ricardo S. Alonso[1,2]

[1] UNIR (International University of La Rioja), Av. de la Paz, 137, Logroño, Spain
carlosluis.fernandezsantana@outlook.es,
{pablo.negre,ricardoserafin.alonso}@unir.net
[2] AIR Institute, Av. Santiago Madrigal, 39, 37003 Salamanca, Spain
{pguilleme,ralonso}@air-institute.com

Abstract. Pediatric appendicitis is one of the most frequent emergency conditions in children, and timely diagnosis is crucial to prevent complications. While ultrasound imaging is commonly used due to its safety and accessibility, its interpretation remains highly operator-dependent, leading to inconsistencies in diagnosis across clinicians and institutions, as reported in prior clinical studies [6]. In this study, the use of Vision Transformers (ViT) is explored for the automated detection of pediatric appendicitis using exclusively real ultrasound images. A Swin Transformer model ("swin tiny patch4 window7 224") is fine-tuned on a curated dataset of abdominal ultrasound images from pediatric patients at the Children's Hospital St. Hedwig (Germany). The model achieves an accuracy of 0.89, along with precision, recall, and F1-score all at 0.89, significantly outperforming the average diagnostic accuracy of initial clinical assessments reported in the dataset (approximately 0.68). These results suggest that ViT architectures can effectively capture complex visual patterns in medical imaging and offer reliable support for clinical decision-making in pediatric appendicitis diagnosis.

Keywords: Pediatric Appendicitis Classification · Swin Transformer · Vision Transformers · Ultrasound Imaging · Medical Image Analysis

1 Introduction

Pediatric appendicitis is one of the most common abdominal emergencies in children and adolescents, with an estimated incidence of 87 cases per 100,000 children per year [10]. Early and accurate diagnosis is essential to prevent complications such as perforation, peritonitis, and unnecessary surgical procedures [1]. Among diagnostic methods, ultrasound imaging is widely adopted due to its non-invasive nature, broad availability, and cost-effectiveness. However, its diagnostic reliability is highly dependent on the operator's skill level and the quality of the imaging equipment [6]. According to recent studies, this dependency often

M. Rodrigues et al. (Eds.): ISAmI 2025, LNNS 1776, pp. 101–108, 2026.
https://doi.org/10.1007/978-3-032-14138-5_11

results in inconsistent interpretations and a notable rate of diagnostic errors in pediatric appendicitis cases [7].

Traditional diagnostic workflows combine clinical symptoms, laboratory markers, and imaging studies. Nevertheless, the subjectivity and variability in ultrasound interpretation present significant limitations for consistent clinical decision-making. Several studies have highlighted that misdiagnoses can occur due to subtle or ambiguous visual features in ultrasound scans, particularly in early or atypical presentations of appendicitis. In response to these challenges, artificial intelligence (AI) has gained relevance as a tool for supporting medical image interpretation, reducing interobserver variability, and improving diagnostic accuracy in a wide range of clinical applications [8].

In addition to its relevance in the field of e-Health, this approach also aligns with the principles of Ambient Intelligence. By enabling real-time, automated interpretation of medical imaging within clinical environments, the proposed system can support context-aware decision-making and enhance diagnostic workflows, particularly in high-pressure emergency settings.

Among the most recent AI advances, Vision Transformers (ViT) have emerged as a promising alternative to Convolutional Neural Networks (CNNs). Unlike CNNs, which focus on local spatial features using convolutional kernels, ViTs use self-attention mechanisms to model long-range dependencies across an image. This approach enables a more comprehensive understanding of global image context, which is particularly advantageous in ultrasound imaging [3]. This paper investigates the application of Swin Transformers, a hierarchical ViT architecture with shifted window attention, to perform binary classification of real pediatric ultrasound images into appendicitis and non-appendicitis categories. The model is trained and evaluated using exclusively real clinical images, without any synthetic data augmentation, to simulate realistic diagnostic conditions.

This paper is structured as follows: Sect. 2 presents related work in deep learning for medical imaging and the role of Vision Transformers. Section 3 describes the dataset, preprocessing steps, and model architecture. Section 4 presents the experimental setup and classification results. Finally, Sect. 5 provides conclusions and discusses potential future work.

2 Related Work

Over the past decade, deep learning techniques, particularly Convolutional Neural Networks (CNNs), have been widely applied in medical imaging tasks, including disease detection, segmentation, and classification [8]. CNNs are known for their ability to extract local spatial features and have demonstrated performance comparable to expert radiologists in specific tasks. However, their effectiveness is often constrained by the need for large labeled datasets and their limited capacity to model long-range dependencies in image data.

To overcome these limitations, Vision Transformers (ViT) have emerged as a promising alternative. Originally introduced for natural language processing, the Transformer architecture was later adapted to computer vision tasks by Dosovitskiy et al. [3]. ViT divides images into fixed-size patches and applies self-attention mechanisms to learn relationships between distant regions in the image. This global context modeling capability enables more comprehensive feature extraction, especially in complex modalities such as ultrasound imaging.

Recent studies have explored the use of ViTs in medical diagnostics, showing competitive or superior performance to CNN-based models. For example, Chen et al. [2] demonstrated that ViTs can effectively classify ultrasound images by capturing subtle texture differences and spatial patterns that CNNs may overlook. These properties are especially relevant in applications like pediatric appendicitis, where distinguishing features may be small, diffuse, or context-dependent.

Despite their advantages, the application of ViT models in clinical practice remains limited, in part due to data scarcity and computational requirements. Most studies have relied on large, pre-augmented datasets or the integration of synthetic images to improve performance. However, there is a growing interest in assessing ViT performance using only real clinical data, to evaluate their potential in realistic, data-constrained environments, such as pediatric emergency diagnosis.

This work contributes to this emerging direction by evaluating the performance of a Swin Transformer model trained exclusively on real ultrasound images for the diagnosis of pediatric appendicitis, without relying on synthetic augmentation techniques.

3 Materials and Methods

This Section describes the dataset used in the study, the preprocessing steps applied to the ultrasound images, the Vision Transformer architecture implemented, and the training configuration used to fine-tune the model. All experiments were conducted exclusively on real ultrasound data, without the use of synthetic images.

3.1 Dataset

The dataset used in this study is the *Pediatric Appendicitis Dataset*, which contains grayscale ultrasound images and structured clinical data collected between 2016 and 2021 at the Children's Hospital St. Hedwig in Regensburg, Germany. The full dataset includes both the presumptive clinical diagnosis (based on initial patient assessment) and the final confirmed diagnosis (following surgical findings or imaging-confirmed resolution). For this study, only the confirmed diagnosis was used to train and evaluate the classification model.

A total of 2,677 real ultrasound images were used: 1,339 from patients with confirmed appendicitis and 1,338 from patients without appendicitis. The dataset was split in a stratified manner to preserve class balance: 80% of the images were assigned to the training set, and 10% each to validation and test sets. Table 1 presents the dataset composition.

Table 1. Distribution of real ultrasound images by subset and diagnosis.

Subset	Total Images	Appendicitis	No Appendicitis
Training Set (80%)	2,141	1,072	1,069
Validation Set (10%)	268	134	134
Test Set (10%)	268	133	135
Total	2,677	1,339	1,338

A relevant feature of the dataset is the availability of both the presumptive and final diagnoses. This allows for a clinical performance analysis of the initial human diagnostic process. Table 2 shows the comparison between presumptive and confirmed diagnoses using standard classification metrics. The results highlight a high recall (0.98) for appendicitis, indicating that most true cases were identified in the initial clinical assessment, but with a relatively low precision (0.65), reflecting a significant rate of false positives. These results reinforce the need for automated diagnostic support systems.

Table 2. Diagnostic Classification Report: Presumptive vs. Confirmed Diagnosis.

	Precision	Recall	F1-Score	Accuracy	Support
No Appendicitis	0.91	0.25	0.39	–	312
Appendicitis	0.65	0.98	0.78	–	443
Macro Avg	0.78	0.62	0.59	0.68	755
Weighted Avg	0.76	0.68	0.62	–	755

3.2 Preprocessing

All images were initially stored in BMP format. They were converted to PNG format to ensure compatibility with modern deep learning frameworks. Grayscale normalization was performed using histogram equalization followed by Z-score normalization. Each image was resized to 224 × 224 pixels to match the input dimensions expected by the model.

Only the confirmed diagnosis was linked to each image for supervised training. The training set was augmented using random rotations, horizontal flips, and adjustments to brightness and contrast to improve model generalization and reduce overfitting.

3.3 Model Architecture

The classification model used in this study is based on the **Swin Transformer**, a hierarchical Vision Transformer architecture introduced by Liu et al. [9]. Swin

Transformer was designed to overcome key limitations of standard ViT models, particularly their fixed-scale representations and high computational cost. By introducing a hierarchical structure and shifted window-based self-attention, the Swin Transformer enables efficient modeling of both local and global features in high-resolution images.

In this study, the specific model variant used is "swin tiny patch4 window7 224". This model divides the input image into non-overlapping patches of size 4×4 pixels, applies attention within 7×7 windows, and uses a patch merging mechanism to progressively reduce spatial dimensions. It is pretrained on ImageNet-1K and fine-tuned for binary classification on the ultrasound dataset.

Model selection was based on the best validation accuracy. Final evaluation was performed on the holdout test set using standard classification metrics: accuracy, precision, recall, and F1-score. These results are presented in the following Section.

4 Experiments and Results

This section presents the evaluation results of the Swin Transformer model trained exclusively on real ultrasound images for binary classification of pediatric appendicitis. The objective is to assess the model's diagnostic performance in a realistic, clinically relevant setting using only real-world data, without the inclusion of synthetic images or augmentation techniques beyond those applied during training.

The model was initialized with pretrained weights from ImageNet-1K and fine-tuned using the real ultrasound dataset. The training process was configured as follows:

- **Optimizer:** AdamW
- **Learning Rate:** 1×10^{-4}
- **Weight Decay:** 1×10^{-5}
- **Batch Size:** 32 (with mixed-precision training)
- **Epochs:** 100 (with early stopping)
- **Loss Function:** Binary cross-entropy
- **Hardware:** Google Colab environment with an NVIDIA A100 GPU

While training was performed on an NVIDIA A100 GPU in a cloud environment, the final model can be exported and run efficiently on modern CPUs or mid-range GPUs for inference. This makes it suitable for deployment in hospital settings with limited computational infrastructure, as the Swin Transformer architecture is relatively lightweight during inference compared to other deep learning models.

Classification Performance and Confusion Matrix

The classification report for the test set is shown in Table 3. The model achieves an overall accuracy of 0.89, with precision, recall, and F1-score values of 0.89

Table 3. Classification report for the Swin Transformer model.

	Precision	Recall	F1-Score	Accuracy	Support
No Appendicitis	0.89	0.89	0.89	–	135
Appendicitis	0.89	0.89	0.89	–	134
Macro Avg	0.89	0.89	0.89	0.89	269
Weighted Avg	0.89	0.89	0.89	–	269

for both classes. These values indicate a highly balanced performance with no observable bias toward either class.

To further analyze prediction errors, Table 4 presents the confusion matrix in terms of true positives (TP), false positives (FP), true negatives (TN), and false negatives (FN).

Table 4. Confusion matrix of the Swin Transformer model.

	Predicted Appendicitis	Predicted No Appendicitis
Actual Appendicitis	119 (TP)	15 (FN)
Actual No Appendicitis	15 (FP)	120 (TN)

The distribution of errors is symmetrical, indicating a well-calibrated model. From a clinical perspective, false negatives—missed cases of appendicitis—carry higher risk due to potential treatment delays. Despite this, the false negative rate remains low (11.2%). Future iterations of the model may enhance sensitivity through threshold adjustments or cost-sensitive learning. Nevertheless, given the potentially severe consequences of missed diagnoses, reducing false negatives remains a clinical priority. Further improvements could involve incorporating multimodal data (e.g., laboratory results or symptoms), fine-tuning decision thresholds, or applying ensemble models to increase diagnostic sensitivity without compromising specificity.

Clinical Comparison and Implications

When compared to the initial clinical assessment (presumptive diagnosis), which achieved an average accuracy of approximately 0.68 and a precision of 0.65 for appendicitis, the Swin Transformer shows significant improvement across all metrics. This suggests that automated diagnostic models based on Vision Transformers can reduce the high rate of false positives commonly associated with initial physician assessments.

The elevated false positive rate observed in expert clinical assessments is likely a consequence of prioritizing patient safety. In pediatric emergency settings, the cost of missing a true appendicitis case is often considered higher than performing an unnecessary intervention. As a result, clinicians may favor over-diagnosis to minimize the risk of life-threatening complications, which leads to lower precision but high recall in human decisions.

Moreover, the model achieves robust performance using only real ultrasound data, with no reliance on synthetic augmentation. These findings support the practical feasibility of integrating ViT-based diagnostic tools into real-world clinical workflows, especially in resource-limited environments where consistent diagnostic expertise may not always be available.

5 Conclusions and Future Work

This study explores the application of Vision Transformers, specifically the Swin Transformer architecture, for the automated diagnosis of pediatric appendicitis using real ultrasound images. The model was trained exclusively on real clinical data and evaluated in a realistic test setting, achieving an accuracy of 89% along with balanced precision, recall, and F1-scores across both classes.

The results confirm that Vision Transformers are capable of capturing complex spatial patterns in ultrasound imaging without the need for large-scale synthetic data. Unlike traditional CNN-based approaches or operator-dependent visual assessments, the Swin Transformer provides consistent diagnostic performance and demonstrates potential for integration into clinical workflows as a decision-support tool.

By achieving strong results in a data-limited environment, the proposed approach supports the viability of ViT-based methods in real-world medical applications, particularly in pediatric emergency contexts where rapid and accurate diagnosis is critical.

Future work may explore several directions. One promising line is the incorporation of synthetic ultrasound images generated via Generative Adversarial Networks (GANs) [5], which could enhance the model's performance by increasing data variability, as demonstrated in previous work involving CNNs [4]. Another important direction involves integrating clinical variables—such as symptoms and laboratory results—into a multimodal diagnostic framework that combines imaging and structured data. Additionally, expanding the model's scope to detect other abdominal conditions, validating it on multicenter datasets to assess generalizability, and exploring interpretability methods to visualize the features influencing predictions, all represent valuable paths for future investigation. Since the dataset was collected from a single medical center, this may limit the model's generalizability to different populations, ultrasound equipment, or imaging protocols. Validating the model across institutions with varied demographics and scanning conditions would be essential to confirm its broader clinical applicability. In particular, attention heatmaps or gradient-based approaches (such as Grad-CAM adapted for ViT) could be explored to highlight the regions of the ultrasound image most relevant to the model's decisions. This added transparency would improve clinical trust and support safer deployment in real-world scenarios.

Acknowledgements. This research has been supported by the Instituto de Competitividad Empresarial de Castilla y León, funded under the "Subvenciones para la realización de Proyectos de I+D; LINEA: 2023 PROYECTOS DE I+D COLABORACIÓN ENTRE CCTT-EMPRESAS" program cofunded by the European Regional Development Fund as part of project "BIOTECARE Avance tecnolgico basado en el dato para la optimizacin de procesos biotecnolgicos, sanitarios y de cuidados" (CCTT6/23/SA/0003).

References

1. Behrman, R.E., Kliegman, R.M., Jenson, H.B.: Nelson Textbook of Pediatrics. Elsevier, Philadelphia, USA (2019)
2. Chen, T.J., Luo, Z., Zhao, H., Wang, X., Yu, S.: Vision transformers in medical imaging: advancements and applications. Med. Image Anal. (2021)
3. Dosovitskiy, A., et al.: An image is worth 16 × 16 words: transformers for image recognition at scale. In: International Conference on Learning Representations (ICLR) (2021)
4. Frid-Adar, M., Diamant, I., Klang, E., Amitai, M.M., Goldberger, J., Greenspan, H.: Gan-based synthetic medical image augmentation for increased CNN performance in liver lesion classification. Neurocomputing **321**, 321–331 (2018)
5. Goodfellow, I., et al.: Generative adversarial nets. In: Advances in Neural Information Processing Systems (NeurIPS), pp. 2672–2680 (2014)
6. Hall, P., Brown, R., Davis, E.: Operator variability in ultrasound interpretation: implications for pediatric appendicitis diagnosis. J. Med. Imaging 243–250 (2021)
7. Kim, Y., Park, J., Lee, S.: Diagnostic challenges in pediatric appendicitis: a clinical perspective. J. Pediatr. Surg. 567–572 (2020)
8. Litjens, G., et al.: A survey on deep learning in medical image analysis. Med. Image Anal. **42**, 60–88 (2017)
9. Liu, Z., et al.: Swin transformer: hierarchical vision transformer using shifted windows. arXiv preprint arXiv:2103.14030 (2021)
10. Snyder, M.J., Guthrie, M., Cagle, S.: Acute appendicitis: efficient diagnosis and management. Am. Family Phys. 25–33 (2018)

Clustering-Based Activity Monitoring Framework for Anomaly Detection in Wearable Data

Arturo Martinez-Rodrigo[1(✉)], Roberto Sanchez-Reolid[1,2], Susana Priego-Jiménez[3,4], and Jose Carlos Castillo[5]

[1] Medical Informatics, E-Health, and Advanced Technologies Researching Group (COMETA), University of Castilla-La Mancha, Cuenca, Spain
arturo.martinez@uclm.es

[2] Department of Computer Systems, University of Castilla-La Mancha, Ciudad Real 13071, Spain

[3] Hospital Virgen de la Luz, 16002 Cuenca, Spain

[4] Health and Social Research Center, Age-ABC Researching Group, University of Castilla-La Mancha, Cuenca, Spain

[5] Systems Automation and Engineering Department, Carlos III University of Madrid, Madrid, Spain

Abstract. Continuous health monitoring using wearable devices has emerged as a promising approach to detecting early signs of physiological and daily activity anomalies in older adults living independently. However, traditional threshold-based methods often fail to account for individual variability, leading to excessive false positives or missed critical events. This study proposes an unsupervised anomaly detection framework that personalizes activity monitoring by clustering based on K-means to consider individual behavioral patterns. Using one-minute time window data from wearable sensors, we establish baseline activity profiles over seven days and detect deviations indicative of potential health risks. Four primary anomaly types were identified: nocturnal restlessness, exertion without movement, prolonged inactivity, and gradual reduction in activity. The system effectively localized these anomalies within a latent feature space using a non-linear dimensionality reduction technique, thus demonstrating its ability to differentiate between distinct deviation patterns.

Keywords: Unsupervised Learning · Wearable Devices · Anomaly Detection · Elderly Monitoring · Activity Clustering

1 Introduction

The global demographic shift towards an aging population has led to a significant increase in older adults living independently. According to recent literature, a substantial proportion of individuals aged 60 and above reside alone,

M. Rodrigues et al. (Eds.): ISAmI 2025, LNNS 1776, pp. 109–118, 2026.
https://doi.org/10.1007/978-3-032-14138-5_12

with this trend being more pronounced among women over 75 [2,17]. This living arrangement poses unique challenges, as aging can affect the ability to maintain consistent daily activity routines due to factors such as decreasing health, reduced mobility, cognitive impairments, or prolonged fatigue [7]. Traditional health monitoring methods, which often rely on infrequent medical visits or subjective caregiver observations, may not effectively capture subtle yet significant changes in an individual's activity pattern. In this context, continuous monitoring of variables like heart rate, physical activity levels, and sleep patterns rises as a means to provide a more accurate assessment of an individual's overall well-being, enabling timely interventions in response to anomalies that could indicate physical or cognitive decline [1]. Recent work in the AmI community has also explored how different types of wearables can support access to smart services for specific populations, such as users with motor impairments, highlighting the importance of context-aware and personalized approaches [18].

Recent literature suggests that an approach tailored to each individual, able to model their unique activity patterns instead of relying on generalized criteria, can more effectively balance sensitivity (identifying truly concerning events) and specificity (minimizing false alarms) [3]. In particular, clustering techniques allow the discovery of different habitual activity dynamics, identifying as anomalies only those deviations that significantly diverge from a person's own reference patterns [11]. This contrasts with fixed rules or thresholds, which fail to account for individual differences. For instance, an activity level deemed concerning for a normally active individual may be normal for a more sedentary person. By profiling each user using unsupervised learning techniques, the system can automatically adapt alert thresholds to each individual's characteristics, improving precision [9].

To address these needs, unsupervised learning is able to identify normal activity patterns and detect deviations without pre-labeled data. [19]. These methods encompass a variety of techniques, ranging from activity clustering to sequence modeling and novelty detection algorithms. For instance, Mori et al. employed activity pattern clustering to identify anomalies based on the duration and timing of various daily activities performed at home [12]. Novák et al. combined self-organizing maps for activity classification with a predictive Markov model, successfully identifying events such as abnormally long periods of inactivity, unexpected presence at atypical hours, or abrupt changes in daily routines [13]. More recently, multivariate outlier detection algorithms and deep learning approaches have been explored, enabling the learning of sensor time-series representations to enhance sensitivity to subtle changes in elderly individuals' habits [14]. However, there is still no absolute consensus on the optimal methodology, and many current systems integrate heuristic rules with unsupervised algorithms, leveraging the strengths of both approaches depending on the context [8]. While these advancements represent significant progress, the development of personalized unsupervised solutions remains a challenging and ongoing research area [8].

This work aims to develop a personalized, unsupervised anomaly detection system for elderly individuals living independently by leveraging wearable activity data from a Fitbit Inspire 3 [5] to construct an adaptive clustering model that dynamically characterizes an individual's routine and identifies deviations that may indicate potential health risks or abnormal routines. Unlike traditional threshold-based methods, this system continuously updates its anomaly detection criteria, reducing false alarms while enhancing sensitivity to critical activity changes.

The rest of the paper is structured as follows: Sect. 2 describes the main steps of the proposed pipeline for anomaly detection from wearable data; Sect. 3 offers a running example to illustrate the operation of the system; Sect. 4 discusses the challenges found during the course of this work and highlights the main conclusions of this work.

2 Methods

This section presents the main steps involved in the analysis of the activity patterns, from data acquisition to anomaly detection.

2.1 Data Collection and Storage Platform

The data collection platform acquires physiological and activity-related data from wearable devices. It connects with Fitbit's API to extract minute-level data on heart rate (BPMs), step count (cadence, measured in steps per minute), and metabolic equivalents (METs), which estimate energy expenditure based on activity intensity. The system performs API requests to Fitbit servers through a Flask-based backend API, which is integrated with a scheduled job execution system (Cometa Server). This setup ensures periodic data acquisition while minimizing API request overhead and avoiding unnecessary server load. The system is designed to fetch and store all available data points, mitigating potential data loss due to API limitations or connectivity issues. Upon retrieval, the collected data undergoes preprocessing, including handling missing values and normalizing timestamps, before being stored in a PostgreSQL database. Figure 1 summarizes the system architecture, illustrating how Cometa Server continuously requests activity data from the Fitbit API server.

2.2 Device Management and Data Visualization

To ensure a scalable and structured data collection process, the system integrates a patient and device management interface, facilitating the registration, linking, and continuous monitoring of Fitbit devices assigned to individual users. Each device is uniquely identified and mapped to a specific patient profile, ensuring accurate data association. Patient information is stored in a dedicated relational database, where demographic and clinical variables (e.g., age, gender, height, weight, etc.) are maintained to support further analysis. The authentication

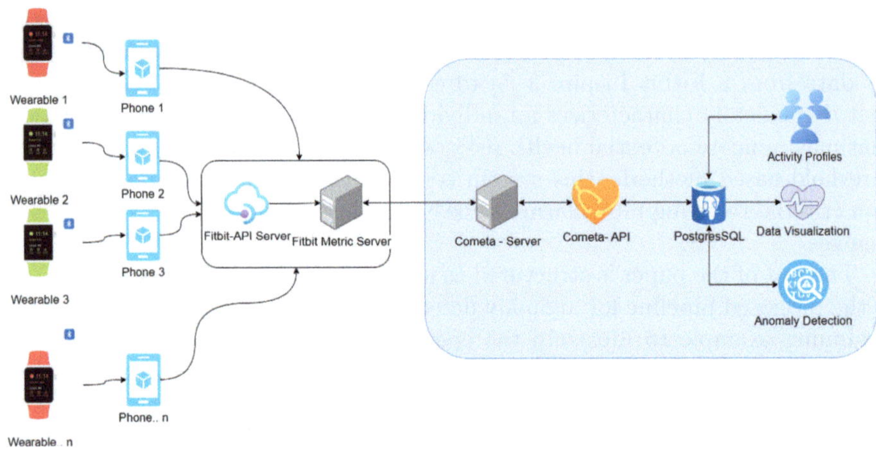

Fig. 1. Architecture for Fitbit data acquisition and metric data processing.

process follows Fitbit's OAuth 2.0 protocol, enabling secure and seamless access to personalized health metrics without manual intervention.

Furthermore, to facilitate the exploration and interpretation of collected physiological data, the system incorporates a real-time visualization module based on Grafana, providing an intuitive and interactive dashboard for monitoring key health metrics. This module connects directly to the PostgreSQL database, enabling the retrieval and visualization of both historical and real-time data trends. The dashboard is designed to display granular information on multiple physiological parameters, including heart rate (BPMs), activity levels (walking cadence), and energy expenditure (METs). Users can interactively filter data by patient ID and date range, allowing personalized assessments.

2.3 Determination of Activity Profiles

To determine each user's activity profile, the analysis focuses on three core variables: BPMs, cadence, and METs, with timestamps standardized to a consistent datetime format. The first seven days of recorded data were selected as a baseline period, assuming this initial phase reflects the user's typical routine. This baseline serves as a reference for detecting deviations, which were subsequently analyzed over a one-month period to identify potential anomalies.

To standardize feature contributions and facilitate clustering and anomaly detection, the selected variables were normalized using Z-score, ensuring that differences in magnitude (e.g., HR in beats per minute vs. cadence in steps per minute) did not bias the model. The optimal number of clusters was determined through both the Elbow Method and the Silhouette Score. The Elbow Method assesses within-cluster variance (inertia) across different values of K, identifying the point where adding more clusters yields diminishing returns. Meanwhile, the Silhouette Score quantifies clustering quality by measuring cohesion (how closely grouped the points within a cluster are) and separation (how distinct clusters are from each other). By integrating the Elbow Method to assess intra-cluster variance and the Silhouette Score to evaluate cluster cohesion and separation, the optimal K value for segmenting user activity states was determined.

K-means demonstrated the best segmentation of activity states and was selected for its consistent cluster formation, enabling the classification of minute-by-minute activity data into distinct behavioral states. After training the clustering model for each user using data from the first seven days, all subsequent activity records from the following month were assigned to the nearest cluster based on Euclidean distance to the centroids in the feature space. This approach ensured that each new data point was mapped to a predefined activity state, providing a continuous representation of the user's routine and a reliable baseline for detecting deviations from typical behavior.

Given the multidimensional nature of the feature space, dimensionality reduction techniques were explored to enhance the visualization and interpretation of clustering results. Both Principal Component Analysis (PCA) and t-Distributed Stochastic Neighbor Embedding (t-SNE) were evaluated, with t-SNE generally providing better visual separation of different activity states for each user.

2.4 Anomaly Detection Engine

Once the activity clusters were established for each user, continuous minute-by-minute data streams from wearable devices were processed to monitor key physiological markers and identify irregularities. Four main categories of anomalies were considered. Table 1 summarizes these anomaly types, detailing their detection criteria and potential clinical relevance.

Table 1. Description of the detected anomaly types and their potential clinical relevance.

Anomaly Type	Detection Criteria	Potential Clinical Relevance
Nocturnal Restlessness	Elevated heart rate combined with increased activity (cadence/METs) between 00:00 and 06:00.	May indicate sleep disturbances, anxiety, pain, or nocturnal hypertension.
Exertion Without Movement	Abnormally high heart rate despite low step count and METs.	Could be associated with nervousness, anxiety episodes, or autonomic dysfunction (e.g., tachyarrhythmias).
Prolonged Inactivity	More than 90% of the day spent in a sedentary state.	May reflect fatigue, depressive states, mobility impairments, or an unnoticed health event.
Gradual Reduction in Activity	Continuous decline in movement levels observed over multiple days.	Suggests worsening health, increased fatigue, or early signs of frailty.

Finally, to identify deviations in user activity and physiological patterns, a two-step anomaly detection approach was implemented. First, each new data point from day seven onward, up to one month, was assigned to the closest cluster based on Euclidean distance to the precomputed centroids. Then, a statistical thresholding method was applied, where the values of HR, step count, and METs were evaluated against the mean and standard deviation of the assigned cluster. Specifically, if any of these variables exceeded ± 2 standard deviations from the cluster centroid's average, the observation was flagged as a potential anomaly.

3 Case Study

To facilitate the interpretability of the proposed anomaly detection framework, the results are presented through a case study focusing on a single user, demonstrating how the system identifies activity profiles and subsequently detects anomalies in daily routines.

3.1 User's Activity Profiles

Firstly, each user was monitored for seven days to establish baseline activity patterns, which served as a reference for the anomaly detection engine. Once the data was collected, the optimal number of clusters (activities) was determined by balancing inertia and silhouette score. In this case, the silhouette score peaked at K = 4 with a value of 0.56, while inertia dropped significantly from over 10,000 at K = 2 to below 3,000 at K = 4. These criteria led to selecting K = 4 as the optimal number of clusters for this specific user. This choice aligns with the expectation that different activity states such as sedentary, light, moderate, and vigorous activity should be distinguishable within the dataset. While K = 4 was optimal for this specific user, the number of clusters may vary across individuals depending on their unique activity patterns.

While PCA provided a linear transformation, t-SNE better preserved local behavioral patterns, confirming K-means' effectiveness in grouping activity states for routine deviation detection. Figure 2 shows four distinct activity patterns based on HR, steps, and METs from the first seven days, capturing daily routines and physiological responses.

To classify each cluster, mean and standard deviation analyses were conducted for HR, steps, and METs, revealing four distinct activity states. Thus, Cluster 0 corresponded to light to moderate activity, with an average heart rate of 96.7 ± 10.8 bpm, 32.2 ± 18.7 steps, and 42.9 ± 9.0 METs, suggesting intermittent movement such as walking or household tasks. Cluster 1 represented rest or sleep, with the lowest values across all parameters: 64.5 ± 5.3 bpm, nearly zero steps (0.1 ± 0.8), and 10.4 ± 1.4 METs. Cluster 2 reflected vigorous activity, characterized by the highest heart rate (109.5 ± 15.7 bpm), high step count (98.0 ± 21.7), and elevated METs (69.4 ± 12.2), indicating exercise or intense movement. Finally, Cluster 3 suggested sedentary behavior, with moderate HR (81.3 ± 6.3 bpm), low step count (1.6 ± 4.2), and low METs (14.9 ± 5.7), consistent with sitting or passive activities.

3.2 Detection of Anomalous Activity Patterns

The trained anomaly detection model was applied to classify new data points based on deviations from learned activity clusters. Figure 3 presents the distribution of detected anomalies in the latent space generated by t-SNE. Normal activity states from the first seven days are shown in gray, serving as the reference distribution.

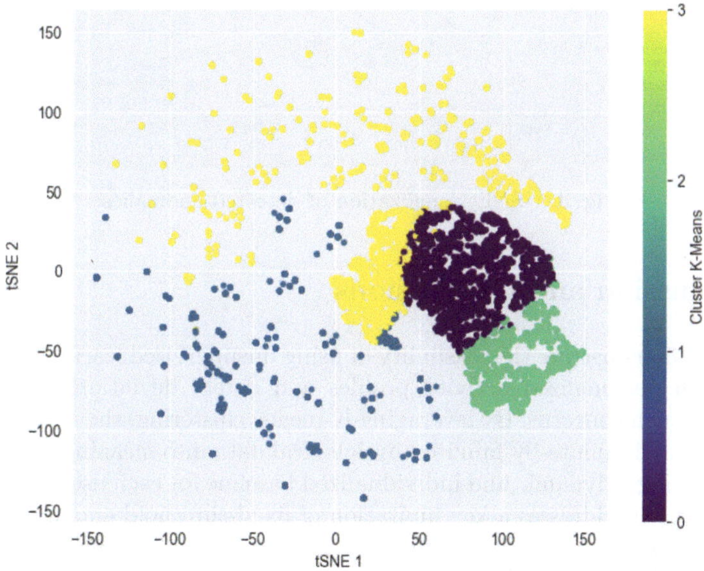

Fig. 2. Visualization of user activity clusters using t-SNE with K-means. Each point represents a minute-level activity record, and colors indicate different activity states.

From day 8 onwards, a total of 14,181 data points were analyzed, resulting in 347 cases (2.4%) of Nocturnal Restlessness and 13 cases (0.1%) of Exertion Without Movement. Although the first type of anomaly was expected to occur during the night, most cases were concentrated in the early afternoon, particularly around 1–2 p.m., which may reflect irregular rest or napping behavior. Notably, instances of Nocturnal Restlessness (blue circles with white edges) appear clustered within a distinct subregion of the latent t-SNE space, suggesting that these events occur within specific behavioral contexts. This aligns with their definition, as such disturbances deviate from standard rest patterns while remaining close to sedentary or low-intensity states. Similarly, Exertion Without Movement events (red crosses) are located in another compact area of the activity space, reinforcing the hypothesis that these anomalies arise from internal physiological responses rather than external movement. Quantitatively, these events exhibited heart rate values exceeding the user's baseline for low-activity clusters by an average of 3.66 standard deviations.

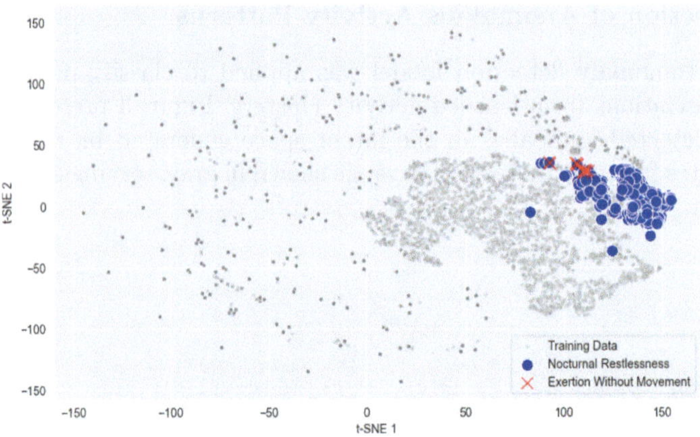

Fig. 3. t-SNE visualization of detected anomalies.

4 Discussion and Conclusions

The results demonstrate the feasibility of using unsupervised learning techniques to establish personalized activity profiles and detect deviations indicative of potential health concerns. By leveraging K-means clustering, the system successfully segmented minute-by-minute physiological data into meaningful behavioral states, allowing a dynamic and individualized baseline for each user. This personalized approach addresses a key limitation of fixed-threshold anomaly detection methods, which often fail to account for inter-individual variability. Moreover, the visualization of anomalies within the latent space revealed that specific deviations, such as nocturnal restlessness and exertion without movement, consistently mapped to distinct subregions. This structured localization supports the effectiveness of the clustering-based anomaly detection approach, reinforcing the potential of unsupervised models in wearable-based health monitoring.

Despite promising results, challenges remain in interpreting detected anomalies. While deviations from learned activity patterns may signal health issues, not all anomalies warrant intervention. Sensor noise, temporary lifestyle changes, and confounding factors (e.g., social activities, medication) can lead to false positives. This issue is common in wearable-based health monitoring, as prior studies have noted difficulties in distinguishing normal fluctuations from clinically relevant deviations [10]. For instance, nocturnal restlessness has been linked to sleep disorders, anxiety, pain, or hypertension, all affecting metabolic and cognitive health [4]. Similarly, prolonged inactivity increases cardiovascular and metabolic risk, stressing the need to differentiate temporary sedentarism from clinically significant patterns [15]. Effort without movement, marked by elevated heart rate without physical activity, may reflect stress, anxiety, or autonomic dysfunctions such as tachyarrhythmias [16]. These cases illustrate that anomaly interpretation demands contextual awareness to reduce false alarms. To address this, future versions should incorporate contextual inputs (e.g., symptoms, medications, environment) and adaptive mechanisms to adjust thresholds based on long-term trends, aligning with the goals of Ambient Assisted Living systems.

Future work should aim to improve anomaly specificity while preserving sensitivity, potentially by integrating contextual factors such as user-reported symptoms or environmental data. Adaptive learning mechanisms could also enhance system robustness as users' routines evolve. Although our findings support the utility of clustering-based detection, validation on a broader and more diverse population is needed to assess generalizability and scalability. In particular, the current evaluation is limited to a single user, and the number of clusters (K) may vary depending on individual patterns. Moreover, due to the unsupervised and personalized nature of the method, standard performance metrics such as accuracy or AUC cannot be computed without ground-truth labels. Future studies incorporating expert-annotated or clinically validated anomalies will be required to enable such evaluations. Despite the promise of continuous monitoring for early detection, as shown in prior studies on infections or arrhythmias [6], the risk of false positives and the balance between sensitivity and specificity remain key challenges.

In conclusion, this study introduces an unsupervised anomaly detection framework for monitoring older adults via wearable devices, overcoming fixed-threshold limitations through personalized clustering. By modeling individual activity patterns with K-means and detecting deviations, the system identifies potential health anomalies like nocturnal restlessness or exertion without movement, among other anomalies. The structured localization of these anomalies validates the approach's effectiveness. However, challenges persist in reducing false positives and improving contextual interpretation of detected deviations.

Acknowledgements. This research has received financial support from several sources: TED2021-130935B-I00, funded by the Spanish Government in conjunction with the European Regional Development Fund (EU) jointly with SBPLY/21/180501/000186, SBPLY/21/180225/000080 provided by the Junta de Comunidades de Castilla-La Mancha, Spain and the European Regional Development Fund (EU); 2024-TRAN-36529 Vascular Applications and Software CompUting NETwork funded by the University of Castilla-La Mancha, PID2021-123941OA-I00, funded by Agencia Estatal de Investigación (AEI), Spanish Ministerio de Ciencia e Innovación.

References

1. Bate, G.L., et al.: The role of wearable sensors to monitor physical activity and sleep patterns in older adult inpatients: a structured review. Sensors (Basel, Switzerland) **23** (2023)
2. Bergh, M., Raval, A., Travers, J.: Supporting older adults who live alone: understanding their unmet needs using NHATS survey data. Innov. Aging **7**, 693–694 (2023)
3. Bijlani, N., Nilforooshan, R., Kouchaki, S.: An unsupervised data-driven anomaly detection approach for adverse health conditions in people living with dementia: cohort study. JMIR Aging **5**(3), e38211 (2022)

4. Bonnet, M.H., Arand, D.: Heart rate variability in insomniacs and matched normal sleepers. Biopsychosocial Sci. Med. **60**(5), 610–615 (1998)
5. Fitbit: Fitbit inspire 3 (2024). https://store.google.com/en/product/fitbit_inspire_3. Accessed Mar 2025
6. Fried, L.P., et al.: Frailty in older adults: evidence for a phenotype. J. Gerontol. A Biol. Sci. Med. Sci. **56**(3), M146–M157 (2001)
7. Gomes, E.S.A., Ramsey, K., Rojer, A., Reijnierse, E., Maier, A.: The association of objectively measured physical activity and sedentary behavior with (instrumental) activities of daily living in community-dwelling older adults: a systematic review. Clin. Interv. Aging **16**, 1877–1915 (2021)
8. Khan, A.Q., Jaouhari, S.E., Tamani, N., Mroueh, L.: Knowledge-based anomaly detection: survey, challenges, and future directions. Eng. Appl. Artif. Intell. **136**, 108996 (2024)
9. Kristijansson, A.M., Aegisson, T.: Survey on technique and user profiling in unsupervised machine learning method. J. Mach. Comput. (2022)
10. Lee, P., Burstyn, I.: Identification of confounder in epidemiologic data contaminated by measurement error in covariates. BMC Med. Res. Methodol. **16** (2016). https://doi.org/10.1186/s12874-016-0159-6
11. Li, J., Izakian, H., Pedrycz, W., Jamal, I.: Clustering-based anomaly detection in multivariate time series data. Appl. Soft Comput. **100**, 106919 (2021)
12. Mori, T., Fujii, A., Shimosaka, M., Noguchi, H., Sato, T.: Typical behavior patterns extraction and anomaly detection algorithm based on accumulated home sensor data. Future Gener. Commun. Networking (FGCN 2007) **2**, 12–18 (2007)
13. Novák, M., Biňas, M., Jakab, F.: Unobtrusive anomaly detection in presence of elderly in a smart-home environment. In: 2012 ELEKTRO, pp. 341–344 (2012)
14. Palermo, F., et al.: Designing a clinically applicable deep recurrent model to identify neuropsychiatric symptoms in people living with dementia using in-home monitoring data. arXiv:2110.09868 (2021)
15. Rezende, L.F.M.D., Rey-López, J.P., Matsudo, V.K.R., Luiz, O.D.C.: Sedentary behavior and health outcomes among older adults: a systematic review. BMC Public Health **14**, 1–9 (2014)
16. Taskasaplidis, G., Fotiadis, D.A., Bamidis, P.D.: Review of stress detection methods using wearable sensors. IEEE Access **12**, 38219–38246 (2024)
17. Torres, J.: Living alone in late life. Innov. Aging **5**, 476–477 (2021)
18. Ungurean, O.C., Vatavu, R.D.: Users with motor impairments' preferences for smart wearables to access and interact with ambient intelligence applications and services. In: Ambient Intelligence – Software and Applications – 12th International Symposium on Ambient Intelligence, pp. 11–21. Springer International Publishing (2022). https://doi.org/10.1007/978-3-031-06894-2_2
19. Wang, Y., et al.: A survey on ambient sensor-based abnormal behaviour detection for elderly people in healthcare. Electronics (2023)

Author Index

M. Rodrigues et al. (Eds.): ISAmI 2025, LNNS 1776, pp. 119–120, 2026.
https://doi.org/10.1007/978-3-032-14138-5

MIX
Papier aus verantwortungsvollen Quellen
Paper from responsible sources
FSC® C105338

If you have any concerns about our products,
you can contact us on
ProductSafety@springernature.com

In case Publisher is established outside the EU,
the EU authorized representative is:
Springer Nature Customer Service Center GmbH
Europaplatz 3, 69115 Heidelberg, Germany

Printed by Libri Plureos GmbH
in Hamburg, Germany